Publish your book like a pro — even if it's your first time

EASY BOOK

SELF-PUBLISHING

A Step-By-Step Guide with AI Assistance
Ready to Use. Easy to Follow.
Everything You Need to Format, Publish, Promote & Sell Your Book.
With Over 150 Ready-to-Use AI Prompts

This step-by-step guide walks you through every step of the publishing process - from ISBNs to LOC (LCCN) and formatting platform-specific walkthroughs for Amazon KDP, IngramSpark, Barnes & Noble, Apple Books, Google Play, and more.

I0054516

Nikolay Gul
https://www.linkedin.com/in/webdesignerny/

Easy Book Self-Publishing – A Step-By-Step Guide with AI Assistance
With Over 150 Ready-to-Use AI Prompts

If you're holding this book in your hands (or scrolling through it on your screen), congratulations you didn't just buy a book. You unlocked a personal publishing mentor, 150 customizable AI prompts, and a creative partnership designed to help you actually finish and launch your book into the world.

You bought it, borrowed it, or believed in it. **That means you're free to** read it, mark it up, copy and paste prompts, and **turn ideas into income**. Please don't repost or redistribute the whole thing. That kills the magic.

Your support helps keep this book affordable and keeps new ones coming. Want to do something awesome? Ask your library to carry it. You get free access. We get more visibility. Everyone wins. Thanks for being the kind of reader who respects the work.

ISBN (Paperback): 979-8-9927440-1-9
Library of Congress Control Number: 2025906574

Cover Design & Published by Nikolay Gul

"Future-Proof Marketing Press is the author's independent imprint name, used for branding purposes only."

Printed in the United States of America

https://www.linkedin.com/in/webdesignerny/

Easy Book Self-Publishing: A Step-by-Step Guide with AI Assistance

---

"This book is the mentor I wish I had when I started."

Introduction: Easy Self-Publishing Starts from This Book

Most aspiring authors believe they need a big budget, a professional team, or years of experience to publish a book. What they don't realize is this: many of the steps they think they have to outsource—they can actually do themselves. They just haven't had a clear, easy-to-follow place to start. Until now.

You're lucky to be holding that starting point.

Easy Book Self-Publishing: A Step-by-Step Guide with AI Assistance is more than a guide—it's a complete self-publishing toolkit. Whether you're a first-time author or already published, this book gives you something rare: **clarity, confidence, and cost-saving control** over your entire publishing journey.

What You'll Learn (and Actually Use)

- **How to professionally publish** on platforms like Amazon KDP, IngramSpark, Barnes & Noble Press, Apple Books, Google Play Books, and Draft2Digital.

- **Avoid expensive beginner mistakes**—from ISBN confusion to messy formatting, metadata misfires, and marketing missteps.

2

- **Get your book accepted, polished, and seen.** You'll follow expert-tested strategies for book formatting, approval, and lasting visibility.

- **Launch smarter. Relaunch better.** You'll discover techniques used by bestselling authors to drive long-term sales and reviews—without gimmicks or hype.

- **Turn your book into a business.** Learn how to expand your book's message into a personal brand, lead magnet, or full-blown income stream.

AI Tools That Work for You—Not the Other Way Around

You'll gain immediate access to over **150 ready-to-use AI prompts** customized for authors and self-publishers. No fluff—just practical, results-driven tools to:

- Generate irresistible book titles and persuasive subtitles.

- Create optimized Amazon descriptions, categories, and keywords.

- Draft social media posts, review request messages, and launch emails.

- Simulate your exact reader with **AI-powered buyer persona testing**, so you know what to say—and how to say it—for maximum emotional connection.

No more guessing who your reader is. No more "hope marketing." Just precision publishing and smart strategy.

Two Game-Changing Bonus Chapters Inside

1. Crafting Your Unique Writing Style with AI

Learn how to use AI to develop a consistent, authentic tone for your writing that's still 100% *you*. Whether you're working on books, blogs, newsletters, or ads - - your writing will sound human, professional, and recognizable. This approach has helped marketers, entrepreneurs, and creatives develop signature styles that build trust, authority, and a loyal audience.

2. Secrets of Successful Preorder Strategies

Most authors skip preorders. That's a mistake. This bonus chapter reveals how you can **build anticipation, increase algorithm exposure, collect early reviews, and drive more day-one sales—** with minimal effort and no need for a huge audience.

No Empty Promises. Just Practical Wins.

This book won't claim you'll become a bestseller overnight. But it **will** show you how to:

- **Save thousands** by avoiding unnecessary service fees.

- **Cut your publishing time in half.**

- **Feel proud of your book** because you know it's well-formatted, well-positioned, and built to last.

You won't just publish. **You'll publish smarter, faster, and with confidence.**

Let's begin.

Contents

14

Chapter 1: Mastering Self-Publishing – Answers to the Questions Every Author Asks

Mastering Self-Publishing – Your Comprehensive Q&A Guide

Starting your self-publishing journey can be exhilarating yet intimidating. With platforms like Amazon Kindle Direct Publishing (KDP), IngramSpark, Draft2Digital, Google Play Books, and Barnes & Noble Press, **clarity and planning** are essential.

This chapter answers the most frequently asked questions I've encountered in real-life publishing and coaching—enhanced with expert tips, practical steps, and **personal experiences you can learn from**.

You'll also find cross-references to deeper guides in later chapters, so you can jump ahead when needed.

---

1. What Exactly is Self-Publishing?

Self-publishing means you are in charge of every stage of producing and distributing your book. Unlike traditional publishing, you don't rely on a literary agent or publishing house. Instead, you handle writing, editing, formatting, ISBN assignment, cover design, publishing platform selection, pricing, marketing, and sales.

Real-Life Example:
When publishing my first book, *AI-Driven Cybersecurity and High-Tech Marketing*, I initially underestimated the depth of work involved. But taking control of every step gave me complete quality oversight, stronger branding, and direct connections with readers.

Actionable Step:
Use this AI prompt to map your publishing roadmap:

AI Prompt: "List all tasks required for self-publishing a nonfiction book, including writing, editing, ISBN registration, formatting, distribution, marketing, and post-launch updates."

Learn more about ISBNs in **Chapter 2**, formatting in **Chapter 3**, and metadata in **Chapter 4**.

---

2. Can I Really Publish a Professional-Quality Book on My Own?

Absolutely. Thanks to modern tools and platforms, you no longer need a publishing contract to produce a professional-grade book. With smart planning and access to AI or affordable freelancers, even first-time authors can launch confidently.

Real-Life Example:
My first book became a #1 new release on Amazon in its category—without a big budget. I followed clear formatting steps, invested in proper editing, and used my own cover templates built in Canva.

Actionable Step:
Make a checklist of the essentials: editing, formatting, cover design, and distribution.

AI Prompt: "List the top 5 most important skills I need to learn as a first-time self-publisher in 2025."

Chapter 3 covers all formatting strategies. Chapter 14 explains how to build your publishing checklist with a minimal budget.

---

3. Do I Really Need an ISBN?

Yes. The ISBN (International Standard Book Number) is what uniquely identifies your book in global retail and library databases. Without it, your book can't be properly cataloged or sold through major channels like Barnes & Noble, IngramSpark, or many international resellers.

Real-Life Example:
Purchasing my ISBN directly from Bowker allowed me to retain control across platforms, while boosting trust with bookstores and libraries.

Actionable Step:
Buy your ISBN from Bowker.com if you're in the U.S., and assign it consistently across every platform you use.

AI Prompt: "Explain the pros and cons of Amazon's free ISBN vs. purchasing your own from Bowker."

Full ISBN registration walkthrough is in **Chapter 2**. Using your own ISBN is also essential if you plan to publish wide (see **Chapter 5**).

---

4. How Should I Distribute My Book?
You have two major strategies:

1. **Amazon-exclusive (KDP Select)** – easier setup, good Kindle reach, but limited to Amazon for 90 days

2. **Wide distribution** – includes IngramSpark, Draft2Digital, Google Play Books, Barnes & Noble Press, and Apple Books

Real-Life Example:
After publishing on Amazon, I used IngramSpark to enable global distribution. My book appeared on Barnes & Noble and several international retailers the next day. However, when I later tried to publish directly on B&N Press, I was blocked because the ISBN was already "in use" by IngramSpark. Lesson learned: **platform order matters.**

Pro Strategy:

- First, publish your eBook and paperback on Amazon KDP

- Then, publish on Barnes & Noble Press (if you want more control)

- Then activate global distribution through IngramSpark

This gives you flexibility over pricing, metadata, and promotional options.

If you prefer simplicity and don't mind lower royalties or return risks, you can activate Amazon's Global Distribution from the beginning.

AI Prompt: "Create a smart multi-platform publishing order for maximum control over pricing and metadata."

Publisher Tip:
*If you activate Amazon's Expanded Distribution at the start, your book may appear on other retailers like Barnes & Noble—but you'll lose pricing control and earn lower royalties (sometimes under 40%). You may also be responsible for the **cost of returned books**, depending on Amazon's agreements with their partners.*

*Always read the fine print and decide whether you want **more reach with less control**, or **more control with a bit more work**.*

Distribution pros and cons are fully explained in **Chapter 5**.

---

5. What If I'm Not Tech-Savvy?
You don't need to be a tech expert to self-publish. Today's tools are designed to make the process accessible.

- Use **Kindle Create** or **Atticus** for formatting
- Use **Canva** for cover design
- Use **ChatGPT** to draft descriptions, brainstorm titles, and even structure your chapters

Real-Life Example:
I formatted and published my book without Photoshop or professional software—just using Kindle Create, Canva, and some smart ChatGPT prompts.

Actionable Step:
Start with user-friendly tools like Kindle Create or Atticus
Consider hiring freelancers for tasks you're uncomfortable with

AI Prompt: "Explain the self-publishing process in beginner-friendly language using simple analogies."

Full tech walkthroughs in **Chapter 3** (formatting) and **Chapter 12** (tools).

---

6. What Are the Most Common Self-Publishing Challenges—and How Can I Solve Them?

- **Market Research**: Know your readers and what they're already buying

- **Book Design**: Your cover sells your book—use Canva or hire a pro

- **Editing**: Essential for quality and trust—don't skip it

- **Marketing**: No visibility = no sales. You'll need a plan

- **Pricing**: A smart price can boost exposure and profitability

Real-Life Example:
I initially priced my paperback too high. Sales were slow. I ran a pricing analysis, adjusted to the market average—and saw a sharp increase in daily orders.

Actionable Step:
Use AI tools to study your competitors and refine your pricing strategy.

AI Prompt: "Analyze pricing strategies for bestselling nonfiction books in [your niche], and suggest optimal ranges for mine."

Learn more about market research in **Chapter 13**, editing and cover design in **Chapters 10 & 11**, and pricing in **Chapter 20**.

---

7. What If I Make a Mistake—Can I Fix It Later?

Yes, and this is one of the biggest advantages of self-publishing. You can update your manuscript, fix errors, revise your description, upload a new cover, or even publish a new edition—all without losing your ISBN or reviews (in most cases).

Real-Life Example:
After launch, I noticed a formatting issue that affected the print version. I re-uploaded corrected files and saw positive reviews return soon after.

Actionable Step:
Create a post-launch checklist to review reader feedback and update files as needed.

AI Prompt: "List the types of updates allowed on Amazon KDP and IngramSpark after publishing."

Learn more about book updates, version control, and second editions in **Chapter 21**.

---

8. How Much Does It Cost to Self-Publish a Book?
Costs vary depending on how much you outsource—editing, formatting, cover design, ISBNs, and marketing. Typically, authors spend anywhere from **$500 to $2,500** to launch a professional-quality book. However, with the right strategy, you can dramatically reduce expenses.

Smart Tip:
With AI tools like ChatGPT, free software like Grammarly and Canva, and careful planning, many new authors successfully publish for **under $250 total**.

AI Prompt:
"Create a realistic low-budget self-publishing plan for a nonfiction book under $250 total."
"Generate a detailed budget breakdown for self-publishing, including optional upgrades like audiobook and print editions."

See detailed cost breakdowns in **Chapter 14: Budgeting Smart – Save Big, Publish Better.**

9. Can I Publish Multiple Platforms Simultaneously?

Yes—and in fact, it's highly recommended. Multi-platform publishing increases visibility, accessibility, and long-term sales opportunities.

Real-Life Example:
Publishing on Amazon KDP, B&N Press, IngramSpark, and Google Play helped my book reach multiple audiences at once. Reviews, traffic, and engagement came from different platforms—without extra marketing cost.

Actionable Step:
Plan your rollout sequence to avoid conflicts (e.g., ISBN conflicts or pricing restrictions). Start with KDP and B&N Press using your own ISBN, then add IngramSpark and Draft2Digital for wider reach.

See **Chapter 5: Smart Publishing Platforms – What to Use, When, and Why** for a full strategy.

---

10. What If I Make a Mistake? Can I Update or Fix My Book Later?

Yes. Self-publishing gives you full control—even after launch. You can revise your book files, fix typos, update the cover, tweak your description, or even release a second edition without losing momentum.

One of the greatest advantages of self-publishing is the ability to revise, update, and re-upload your files at any time. You can update your manuscript, cover, description, price, and keywords. Even publish a second edition or expanded version later.

Pro Author Strategy: Treat your book as a living asset. Improve it over time based on feedback and market trends.

Real-Life Example:
A formatting error in my first upload led to some early negative reviews. I corrected the issue, re-uploaded the book, and saw my average rating rebound within a few days.

Actionable Step:
Set a regular schedule to review feedback, improve your content, and keep your book up to date.

AI Prompt:
"What can I legally and technically update on Amazon KDP or IngramSpark after publication?"

Learn how to revise your book smartly in **Chapter 21: Updates, Versions, and Second Editions**.

---

11. How Long Does It Take to Self-Publish a Book?

Depending on your book complexity, publishing can range from a few weeks to several months. Typically, allowing at least 8-12 weeks ensures sufficient time for editing, formatting, and distribution planning.
For first-time nonfiction authors, a realistic timeline is **8 to 12 weeks** from final draft to full distribution—faster if you're well-organized.

Average Timeline:

- **Editing**: 2–3 weeks

- **Formatting**: 1 week

- **Cover Design + ISBN**: 1 week

- **Upload + Approval**: 1–3 days

- **Marketing & Launch Prep**: 2–4 weeks

Real-Life Example:
It took me just under 10 weeks to move from a final manuscript to

global publication, including metadata setup, testing platforms, and preparing launch content.

AI Prompt:
"Create a 90-day self-publishing timeline for a nonfiction author, including weekly milestones."

See **Chapter 23: Publishing Timeline – Plan, Track, and Launch Without Stress** for a full breakdown.

---

12. How Can I Ensure My Book's Quality?

Quality is key. Here's what you can't skip:

- **Professional Editing** – At least use AI + freelance proofreading

- **Striking Cover Design** – Use templates from Canva or hire an experienced designer

- **Flawless Formatting** – Ensure spacing, margins, and chapter layout meet modern standards

Full walkthroughs in **Chapter 3: The Hit – Formatting That Sells**, and **Chapter 11: Covers That Convert**.

---

13. How Do I Avoid the 'Self-Published Look'?

Avoid:

- Bad trim sizes (e.g., too large or small)

- Amateur covers with weak fonts or images

- Inconsistent formatting or missing Table of Contents

- Cheap or pixelated interior layout

Solution: Follow platform specifications, use trim sizes like 6"x9", and use tools like Atticus or Kindle Create.

More layout standards in **Chapter 3** and interior design in **Chapter 12**.

---

14. What Are the Most Effective Marketing Strategies?

- Create a **book website or landing page**
- Leverage **Amazon's A+ Content** and Author Central
- Use **email newsletters** and early reader lists
- Join **LinkedIn and Facebook groups** in your niche
- Submit your book to **free and paid promotion sites**

See **Chapter 22: Book Launches and Visibility** for affordable, step-by-step strategies.

---

15. Should I Enroll in Kindle Unlimited (KU)?

Kindle Unlimited is best for authors writing engaging fiction. Understand the exclusivity and revenue based on pages read, and weigh it against other distribution options.

Only if you write fiction or serialized content. KU requires Amazon exclusivity for 90 days and pays based on pages read—not great for most nonfiction.

See **Chapter 6: Kindle Exclusive vs. Wide Publishing** to compare your options.

---

16. What Are the Latest Trends in Self-Publishing?

- **AI for content creation** (blurbs, outlines, metadata)
- **Digital-first releases** with print-on-demand paperback later
- **Voice-first formats** like audiobooks for nonfiction

- **Hybrid strategies** that combine wide and exclusive rights

More in **Chapter 16: The Future of Publishing – Where Things Are Headed**.

---

17. How Do I Protect My Work?

- Register with the **U.S. Copyright Office**
- Enable **DRM protection** where available
- Add your brand or logo to interior pages
- Track using **Google Alerts** or **ISBN lookup tools**

Full copyright guide in **Chapter 7**.

---

18. What's a Smart Pricing Strategy?

- eBooks: $2.99–$9.99 for 70% royalties on Amazon
- Paperbacks: Price above print cost but competitive within your genre

See calculators and examples in **Chapter 20: Smart Pricing That Works**.

---

19. What Are the Best Promotional Strategies?

- Use **Amazon Ads** or **Meta (Facebook/Instagram)** ads with tight targeting
- Join **book giveaway campaigns**
- Ask for early **reader reviews** from friends or niche audiences
- Offer a free bonus download inside your book to build your list

Step-by-step campaigns in **Chapter 22**.

20. How Can I Maximize Results Beyond Book Sales?

Even if you don't aim to profit from sales alone, your book can help you:

- Build an audience

- Offer paid coaching or workshops

- Sell a related course or workbook

- Promote your website, newsletter, or services

Monetization extensions in **Chapter 24**.

21. What Makes a Self-Published Book Stand Out?

- A benefit-driven subtitle

- A clean, professional cover

- Well-formatted print and digital editions

- A strong, SEO-friendly description

- Honest early reviews

Learn how to position your book in **Chapter 4** and earn trust with readers in **Chapter 13**.

22. What's the Future of Self-Publishing?

Expect more integration of:

- AI-powered publishing tools

- Subscription-based reading platforms

- Global distribution via POD

- Hybrid author services that combine DIY with pro support

Strategic preparation guide in **Chapter 25: Future-Proofing Your Publishing Career**.

Self-publishing isn't just possible, it's personal, powerful, and rewarding.

23. The Author's Shortcut: AI-Powered Solutions to the Most Popular Self-Publishing Questions

The top questions real authors are asking—but in a modern, AI-enhanced and actionable format. These aren't just theoretical questions, they are the exact points that 90% of new authors struggle with *right now*, like:

"Do I really need an ISBN, or can I just publish on Amazon without it?"

AI prompt: *Help me decide if I should buy my own ISBN or use the free one on KDP.*

"How do I get my book into libraries?"

LCCN vs. Bowker Metadata vs. IngramSpark reach.

AI prompt: *Generate a short library pitch based on my book description.*

"What makes a book 'look professional' to readers and reviewers?"

Formatting checklist + cover design red flags.

AI prompt: *Evaluate my book's title and subtitle for first impressions.*

"How do I price my book smartly for both Amazon and retail?"

Real-world pricing psychology and royalty tips.

AI prompt: *Analyze pricing trends in my category and suggest a smart price range.*

"I hate marketing. What's the bare minimum I should do?"

Author brand shortcuts + metadata hacks.

AI prompt: *Create a low-effort book launch plan based on my publishing timeline.*

"What is timeline to publish my book this year?"

Fast-track publishing timeline checklist.

AI prompt: *Create a custom 30-day publishing plan based on my draft status.*

By following the guidance in this chapter and diving deeper into the referenced sections, you'll avoid common mistakes, save hundreds or even thousands of dollars, and experience the unmatched satisfaction of holding your finished book in your hands—created on your terms.

You've already taken the first step.
Now, let's make your book real.

Chapter 2: Mastering Your Book's Audience – Revolutionary Buyer Persona & Market Demand Blueprint with AI

Publishing a book without knowing your exact reader is like setting sail without a compass—you'll drift aimlessly, miss your audience, and waste valuable resources. This chapter introduces a groundbreaking, future-proof method for creating precise buyer personas and accurately gauging market demand using innovative AI-driven strategies. By following these easy-to-personalize AI prompts, you'll uncover exactly who your readers are, what they crave, and how best to deliver your message to captivate and convert.

Why You Need Accurate Buyer Personas

Clearly defined buyer personas:

- Dramatically enhance your marketing precision.

- Improve your book's discoverability on Amazon and beyond.

- Maximize your ROI (Return on Investment) by precisely targeting ads.

- Provide clarity and confidence, allowing authors to craft resonant content that deeply connects with readers.

Step-by-Step Blueprint to Creating Powerful Buyer Personas with AI

Step 1: Clearly Define Your Initial Idea

Begin by articulating the core topic and genre of your book.

AI Prompt: *"Summarize the core topic, genre, and key audience benefits of a nonfiction [or fiction] book about [your topic] in an engaging, clear manner."*

Step 2: Identify Your Ideal Reader Demographics

Your book's success hinges on knowing precisely who your readers are.

AI Prompt: *"Generate a detailed demographic profile of readers who would most benefit from a [genre/topic] book focused on [key problems it solves]. Include age, gender, education, occupation, income level, location, and lifestyle preferences."*

Step 3: Uncover Deep Emotional Drivers and Reader Motivations

Identify what genuinely motivates readers to buy your book.

AI Prompt: *"List emotional triggers and deep motivations of readers seeking solutions provided by a book about [your topic]. Include fears, desires, aspirations, and personal challenges."*

Step 4: Perform an AI-Powered A/B Testing of Buyer Personas

Use AI to objectively evaluate and refine your personas for accuracy.

AI Prompt: *"Create two distinct buyer persona descriptions for a book about [your topic]. Persona A targets [demographic 1], Persona B targets [demographic 2]. Perform a fair, unbiased analysis comparing market size, demand potential, likely engagement, and profitability."*

Step 5: Advanced Persona Validation Through Real-World Simulation

Realistically simulate potential audience engagement and response.

AI Prompt: *"Simulate realistic reader responses for Persona A and Persona B after seeing an ad and book description for a book on [your topic]. Include likely initial interest, emotional reactions, objections, and buying intent."*

Step 6: Measure Market Demand Precisely with AI

Validate your topic's demand potential in your target audience.

AI Prompt: *"Provide a realistic estimation of market demand, audience size, and potential reader growth for a book on [your topic] within [target demographic or market niche]. Include future growth potential and relevant market trends."*

Step 7: Refine and Personalize Your Ultimate Buyer Persona
With clear insights, finalize your optimized buyer persona.

AI Prompt: *"Create an ultimate, optimized buyer persona for my book on [your topic], combining insights on demographics, psychographics, motivations, emotional triggers, and demand data. Present the persona in a clear, engaging narrative format."*

Easy Implementation and Unmatched Benefits

Implementing these steps is straightforward, enjoyable, and highly empowering. This groundbreaking AI-driven process provides:

- **Immediate Clarity**: Understand precisely who your audience is, eliminating uncertainty.

- **Optimized Marketing**: Craft laser-targeted book descriptions, ads, and promotional materials that speak directly to your readers.

- **Increased Sales**: Higher conversion rates from marketing efforts and improved book rankings.

- **Long-term Success**: Consistently identify opportunities for growth and expansion in future projects.

Real-World Success Example

Consider author Emma, who published a self-help book. Using these AI prompts, Emma clearly identified a previously overlooked persona—professional women transitioning careers. Emma adjusted her marketing materials and Amazon keywords accordingly. Sales increased by 230%, and her book rapidly climbed bestseller lists.

Encouragement and Optimism

You already possess the most important assets, meaningful message. Leveraging these innovative AI-powered techniques makes identifying your audience simpler, more enjoyable, and incredibly effective. This isn't just a technical process; it's your key to confidently connecting your wisdom with readers eager to benefit from it.

You're now equipped with an unmatched edge—one that's both future-proof and effortless to personalize.

Your book's audience awaits. Let's go capture their hearts and minds!

Quick Recap & Action Checklist:

- Define your book's initial idea and benefits.

- Identify demographics clearly.

- Discover deep emotional drivers.

- Conduct AI-powered A/B persona tests.

- Simulate realistic audience responses.

- Validate market demand using AI.

- Finalize your ultimate buyer persona.

Now, you're ready—go confidently and share your book with the perfect audience awaiting your wisdom!

Chapter 3 The Step-by-Step Roadmap to Publishing Success

Completing your manuscript is a major milestone—but publishing it smartly is what leads to real success. This chapter walks you through the essential, practical steps of publishing in the optimal order, with time-saving guidance and AI-enhanced decision-making tools that empower you from start to finish.

Step 1: Choosing and Purchasing Your ISBN

What Is an ISBN, and Why It Matters

An ISBN (International Standard Book Number) uniquely identifies your book and format. While Amazon offers free ISBNs, purchasing your own provides full control over your publishing rights and allows you to publish on multiple platforms (like IngramSpark or Apple Books) without restrictions.

Where to Purchase ISBNs

U.S. authors should buy ISBNs through Bowker's MyIdentifiers.com.

Vendor	Cost	Recommended Usage
Bowker	1 ISBN: ~$12510 ISBNs: $295	Best for U.S. Authors
Nielsen UK ISBN	~£89 per ISBN	UK-based authors

Why You Need Multiple ISBNs

You'll need a separate ISBN for each version of your book: paperback, hardcover, eBook, and audiobook. Owning your ISBNs gives you credibility and flexibility to publish.

Book Format	ISBN Required?
Paperback	☑ Yes
Hardcover	☑ Yes
eBook	☑ Yes
Audio-book	☑ Yes

AI Prompt: *"Help me create a publishing plan that estimates how many ISBNs I need for my book and formats."*

Step 2: Registering Your Book with the Library of Congress
What Is an LCCN and Who Needs It

The Library of Congress Control Number (LCCN) is useful for nonfiction and academic authors who want their books available in public and university libraries.

How to Apply for Free

Visit loc.gov/publish/pcn and apply through the PrePub Book Link. After approval, you'll receive your LCCN via email and can include it in your book's copyright page.

AI Prompt: *"Walk me through applying for an LCCN registration for my nonfiction paperback."*

Step 3: Publishing First on Amazon KDP
Why Start with Kindle Direct Publishing (KDP)

KDP provides global reach, a fast publishing dashboard, print-on-demand support, and access to Kindle and paperback markets. It's often the first and best place to publish.

Create an Amazon KDP account

Task	Format	Status
Create KDP account	Amazon KDP platform	☑
Format manuscript	DOCX, PDF (print-ready)	☑
Cover design	JPEG (eBook), PDF (print)	☑
Metadata (title, keywords, categories)	Text form	☑
Pricing and royalties	Select on dashboard	☑

AI Prompt: *"Write a compelling Amazon book description using bullet points for a nonfiction guide about [topic]."*

Step 4: Expanding Distribution via IngramSpark & Beyond

When and Why to Use IngramSpark

Once your book is live on Amazon, use IngramSpark to distribute to indie bookstores, libraries, and global print markets. You must use your own ISBN to publish on IngramSpark.

Avoiding Conflicts Between Platforms

Never use Amazon's free ISBN for IngramSpark. Publishing order matters—start with KDP, then use the same ISBN to upload to IngramSpark after KDP is live.

Action	Recommended ISBN
Amazon-only publishing	Amazon Free ISBN
Multi-platform publishing	Your purchased ISBN

AI Prompt: *"List the steps to publish my paperback on IngramSpark after launching on KDP."*

Step 5: Formatting Essentials
Standard Formatting Guidelines

Element	Recommended Formatting
Trim size	6" x 9"
Font	Times New Roman, 12 pt
Line spacing	1.15–1.5
Margins	Top/Bottom: 0.5", Inside: 0.75"
Paragraph indents	Use styles, not tabs or spaces
Page breaks	Insert after chapters

Common Formatting Mistakes to Avoid

- Don't use tabs for paragraph indents—use paragraph styles
- Insert page breaks after each chapter
- Keep fonts consistent across formats

AI Prompt: *"Format this paragraph in KDP-friendly style with 6x9 layout and clean indenting: [paste paragraph]."*

Step 6: Strategic Publishing Timeline
Launch Timeline Template

Week 1: Finalize manuscript + cover

Week 2: Upload to KDP + soft launch Kindle edition

Week 3: Collect ARC feedback, refine paperback

Week 4: Publish on IngramSpark, update metadata, launch promo ads

Momentum Strategy

Use early reviews and ARC readers to create buzz before your full release. Schedule promotional posts and email blasts to coincide with paperback launch.

AI Prompt: *"Create a 4-week publishing timeline for my nonfiction book that launches first on Kindle, then expands to print."*

Final Thoughts

By following this upgraded, practical roadmap, you'll

- Make smarter decisions,
- Avoid delays, and publish with confidence. With each step, you'll be building a solid publishing foundation that maximizes your reach, maintains professional standards, and brings your message to the world.

eBook vs Paperback Checklist

Feature	eBook (Kindle)	Paperback (KDP)
File Format	DOCX / EPUB	Print-ready PDF
Cover	JPEG (front only)	Full-wrap PDF (front/spine/back)
ISBN	ASIN or Bowker ISBN	Bowker ISBN recommended
Tools	Kindle Create, Vellum	Atticus, Word PDF Export (PDF/A)

You've already written the book—now let's make sure the world can read it.

Chapter 4: Your Book Formatting That Sells, Gets Approved & Feels Premium

Formatting is the one step that separates amateur-looking books from ones that feel professionally published—and it's where most great manuscripts stumble.

Whether you're formatting for Kindle, paperback, or both, poor layout leads to file rejection, negative reviews, and lost sales. This chapter gives you a powerful, step-by-step, cross-platform formatting system used in bestselling books and publishing agencies.

Why formatting matters

Readers might forgive a typo—but not a disorganized or unprofessional layout. Clean formatting improves:

- Reader experience

- Platform approval rates

- Visual trust and credibility

- Sales and reviews

Your content may be brilliant, but formatting is what makes it **accessible, publishable, and powerful.**

Section 1: Universal formatting standards (6x9 nonfiction recommended)

Element	Format/Value
Trim size	6 x 9 inches (industry standard)
Font	Times New Roman 12 pt (or Cambria 11 pt)

Element	Format/Value
Line spacing	1.15 (print), 1.3 (Kindle)
Paragraph indentation	First-line indent: 0.5 inches
Margins	Top/Bottom: 0.5", Inside (Gutter): 0.75", Outside: 0.5"
Text justification	Fully justified
Image resolution (print)	300 DPI minimum
Page breaks	Always insert between chapters, front/back matter
Avoid	Tabs, inconsistent styles, spacebar indents

AI Prompt:

"Format my 6x9 nonfiction manuscript with Times New Roman 12 pt, 1.15 spacing, justified text, and embedded fonts."

Section 2: Front matter checklist

Use the same layout for eBook and print, with adjustments for interactivity (hyperlinks) in eBooks.

Required front matter:

- Title page

- Copyright page (include ISBN, LCCN, edition, year, rights)

- Dedication (optional)

- Table of Contents (page numbers or hyperlinks)

- Foreword or introduction

AI Prompt:
"Generate a clean front matter layout for a nonfiction 6x9 paperback, including LCCN and ISBN."

Step 3: Platform-specific formatting, in publishing order

1. Bowker – ISBN registration (U.S.)
You must register your ISBN at MyIdentifiers.com.

Requirements:

- Title & subtitle

- Format (paperback, eBook, audiobook, etc.)

- ISBN (one per format)

- Author name

- Publisher name (even if self-published)

- Publication date

- Page count

- Audience category

- BISAC codes

- Cover image (JPEG, 300 DPI)

AI Prompt: *"Create Bowker ISBN metadata for a nonfiction paperback and eBook with 200 pages and self-published author info."*

2. Library of Congress – LCCN (U.S. only)
Apply for free via: https://www.loc.gov/publish/pcn

Steps:

- Apply before publishing

- Include ISBN and title page (PDF preferred)

- Receive your LCCN via email

- Add LCCN to your copyright page

AI Prompt: *"Walk me through LCCN registration for my nonfiction paperback."*

3. U.S. Copyright Office – copyright registration
Website: https://www.copyright.gov

Formats accepted: DOCX, PDF, EPUB, RTF
Optional but smart: Upload cover image (JPEG or PDF)
Cost: $45–$65
Timeline: 3–6 months (expedited available)

AI Prompt:
"Generate a copyright-safe PDF with embedded fonts for my nonfiction book."

4. Amazon KDP – Kindle eBook

Feature	Requirement
Interior	DOCX or EPUB
TOC	Must be hyperlinked
Cover	JPEG, min. 2560x1600 px, 300 DPI
Platform Tool	Kindle Previewer for testing

AI Prompt:

"Convert my manuscript to Kindle-ready EPUB with clickable TOC and embedded metadata."

---

5. Amazon KDP – Paperback

Feature   Requirement

Interior file Print-ready PDF, embedded fonts

Trim size   6 x 9 inches

Margins   Top/Bottom: 0.5", Inside: 0.75", Outside: 0.5"

Cover   Full-wrap PDF (spine + back + front), use KDP Cover Calculator

AI Prompt: *"Generate spine width and full wrap cover for 150-page 6x9 paperback on white paper."*

6. Barnes & Noble Press – eBook and Paperback

Format   Interior   Cover File

eBook   EPUB or DOCX JPEG (300 DPI)

Paperback PDF or DOCX   Full wrap PDF or JPEG front

AI Prompt:

"Design a 5.5x8.5 paperback cover for B&N Press with 200 pages and export compliant PDF."

7. IngramSpark – paperback and eBook

Requirement Details

Interior   PDF/X-1a:2001 only

Cover   PDF/X-1a:2001, CMYK, 300 DPI, no ICC

Requirement Details

Tools   Use their templates for exact dimensions

Notes   Your own ISBN is required

AI Prompt:
"Export PDF to PDF/X-1a:2001 for IngramSpark with embedded fonts and CMYK color."

8. Apple Books – eBook only

| Format | EPUB only (must pass EPUBCheck) | | Cover | JPEG or PNG (300 DPI) | | Note | Layout must be responsive (no fixed pages) |

AI Prompt:
"Check my EPUB for Apple Books compatibility using EPUBCheck."

---

9. Google Play Books – eBook or PDF

| Interior | EPUB (preferred), PDF accepted | | Cover | JPEG or TIFF (300 DPI) | | Size limit | Must be under 2GB |

AI Prompt:
"Optimize EPUB for Google Play with size check and responsive layout."

10. Draft2Digital, Kobo, Lulu, Smashwords (if applicable)

All accept EPUB, and most will generate print-on-demand or distribute to additional platforms (Apple, B&N, OverDrive, libraries).

Check each site's specs before submitting.
All require your own ISBN if distributing wide.

Step 4: Tools that simplify formatting

Tool	Best Use
Atticus	Print + eBook formatting (Mac/Windows)
Vellum (Mac only)	Beautiful layout automation
Reedsy Editor	Free, browser-based interior formatter
Kindle Create	Quick Kindle conversion and preview
Canva / InDesign	Cover design with precision
Format Painter	Style consistency in Microsoft Word

Tips: Always preview before uploading

- Save as PDF "best for printing" with fonts embedded

- Test EPUBs with EPUBCheck and Kindle Previewer

Step 5: Avoid these formatting mistakes

- Using Amazon's free ISBN and trying to list on IngramSpark (will get rejected)

- Forgetting to embed fonts in print PDFs

- Mixing font styles or sizes

- Skipping page breaks between chapters

- Low-resolution images or RGB color space for print

- Tabs or spacebar indentations instead of style-based formatting

Step 6: Final thoughts and AI automation tips

You don't need to master InDesign or hire expensive designers. With the right tools and prompts, you can:

- Format once and export everywhere

- Ensure platform compliance

- Create professional-quality print and eBook editions

- Avoid the "self-published look" that turns readers away

Final AI Prompt: *"Checklist of file formats and specs needed to submit a 6x9 nonfiction book to Amazon KDP, IngramSpark, and B&N."*

This one chapter isn't just a formatting guide. It's a publishing playbook.

Format	ISBN Required?	Notes
Kindle eBook	✕ (uses ASIN)	Unless "Wide" distribution (Apple, Kobo, etc.)
EPUB eBook	✓	Required if sold outside Amazon
PDF eBook	✓	Required if listed on platforms like B&N, Ingram
Paperback	✓	One ISBN per trim size & format
Hardcover	✓	Optional, but treated as separate product

Use it every time you upload a book. Copy/paste the prompts.
Build your own template. Avoid the common traps.
And watch your book go from "uploaded" to "unforgettable."

Chapter 5. The Ultimate Image and Manuscript Formatting Guide – Your All-in-One Reference

Image Quality & File Preparation Checklist.

"This chapter expands on formatting with deeper insights into image types, cover specs, and platform-specific file compatibility."

Welcome to your indispensable resource for mastering book formatting. Proper formatting ensures your book looks professional, gets approved quickly, and provides your readers with an optimal reading experience. This chapter consolidates everything you need, complete with easy-reference tables, practical tips, and customizable AI prompts.

Manuscript Standards & General Guidelines

Proper manuscript formatting is crucial to acceptance and readability:

- **Trim Size:** Standard recommended size: **6" x 9"**.

- **Margins:**

Top & Bottom: 0.5"

Inner (Gutter): 0.75"

Outer: 0.5"

- **Fonts:** Times New Roman, Garamond (Recommended)

- **Font Size:** 12 points

- **Line Spacing:** 1.15 to 1.25 spacing recommended

- **Paragraphs:** First line indentation: 0.25"

AI Prompt for Formatting Check: *"Review the manuscript provided. Highlight formatting issues according to standard publishing* *guidelines for margins, fonts, and spacing. Suggest precise corrections for immediate compliance."*

Images: Types, Formats, and Best Practices

Use these image guidelines to ensure clear, professional results:

- **Resolution:** 300 DPI (pixels/inch)

- **Color Modes:** CMYK for print, RGB for digital (eBook)

- **Formats Explained:**

JPEG: Front cover, eBooks, photos

PNG: Transparent backgrounds, logos

TIFF: High-quality print images (optional)

PDF: Recommended full-wrap cover, interior print-ready

Image Use	Recommended Size (px @300DPI)	Recommended Formats
eBook Cover	1800 x 2700 (6"x9")	JPEG, PNG
Print Front Cover	1800 x 2700 (6"x9")	JPEG, PDF
Full Cover Wrap	3750 x 2775 (12.5"x9.25")	PDF
Interior Images	Max width 1500 pixels	JPEG, PNG

AI Prompt for Image Optimization: *"Analyze my uploaded image files. Recommend optimal formats, DPI settings, and color profiles based on intended publishing platforms. Highlight required adjustments."*

Cover Images & Interior Layout

Your book cover and interior layout speak volumes:

- **Front Cover:** Must be clear, attractive, legible

- **Spine:** Clearly readable text (book title, author's name)

- **Back Cover:** Concise book description, ISBN, barcode

Quick Reference Table: Platform-Specific Cover Guidelines

Platform	Front Only	Full Wrap	Manuscript
Amazon KDP	JPEG	PDF	EPUB/PDF
IngramSpark	JPEG	PDF/X-1a:2001	EPUB/PDF
Barnes & Noble	JPEG/PDF	PDF	EPUB/PDF
Apple Books	JPEG/PNG	N/A	EPUB
Google Play	JPEG/TIFF	N/A	EPUB/PDF

AI Prompt for Cover Check: *"Evaluate the provided cover files for compatibility across Amazon KDP, IngramSpark, Barnes & Noble, Apple Books, and Google Play. Provide a detailed report on suitability and required improvements."*

Quick Fixes & Troubleshooting

Common formatting pitfalls and quick solutions:

Issue	Common Cause	Quick Solution
File too large (>2GB)	Excessive DPI (>300)	Reduce resolution to 300 DPI, compress as JPEG/PDF
Manuscript rejection	Non-embedded fonts	Re-save PDF with embedded fonts
Blurry cover	Low DPI or incorrect size	Re-export at 300 DPI with proper dimensions

AI Prompt for Troubleshooting: *"The manuscript/cover is rejected due to formatting issues. Analyze the file provided, diagnose the exact problems, and suggest clear step-by-step corrections."*

AI-Assisted Formatting Tools

Harness AI to automate tedious tasks:

AI Format Checkers: AI Prompt: *"Perform a thorough formatting check for the manuscript against Amazon KDP/IngramSpark standards."*

Cover Design Assistants: AI Prompt: *"Suggest improvements to my cover design to meet professional standards for readability, clarity, and appeal."*

Customizable AI Prompt Reference

Use these quick prompts for immediate assistance:

Quick Formatting Review: *"Perform a rapid formatting analysis of my manuscript for instant publication readiness."*

Image Quality Check: *"Review image files provided for compliance with professional standards (resolution, size, color)."*

Platform Compatibility Check: *"Evaluate provided book files for cross-platform publishing suitability. Provide actionable feedback."*

With this guide and these powerful AI prompts, formatting your book becomes effortless. Enjoy the peace of mind and confidence in knowing your publication is professionally polished, perfectly formatted, and ready to impress your readers.

Stay focused. Stay precise. Keep creating.

Chapter 6: The Definitive Step-by-Step Guide to ISBN Registration with Bowker for Self-Publishers

This chapter is your hands-on, tested, real-world guide to registering ISBNs using Bowker (MyIdentifiers.com). Whether you're publishing your first book or adding to your author catalog, this is the system that was used to publish Easy Book Self-Publishing (ISBN: 9798992744019) and AI-Driven Cybersecurity and High-Tech Marketing (ISBN: 9798218612481).

You'll get everything you need:
- A simple ISBN breakdown
- Proven best practices
- Time-saving AI prompts
- Metadata walkthrough based on real Bowker entries

Step 1: What Is an ISBN and Why It Matters
An ISBN (International Standard Book Number) is a unique 13-digit code that identifies a book's title, format, and edition. You need a separate ISBN for every version of your book:
- Paperback = 1 ISBN
- Hardcover = 1 ISBN
- eBook (EPUB/PDF/etc.) = 1 ISBN

Important:
You don't need an ISBN for Kindle-exclusive eBooks if you accept Amazon's free ASIN. But if you plan to sell on Apple Books, Google Play, IngramSpark, or libraries, you must provide your own ISBN—even for eBooks.

Step 2: Buy ISBNs from Bowker (MyIdentifiers.com)
Bowker is the only official ISBN agency in the U.S. Do not buy ISBNs from resellers or shady services.

How to Set Up Your Bowker Account:
- Go to https://www.myidentifiers.com

- Create a new account using a professional email
- Save your login details securely

Pricing (as of 2025):
- 1 ISBN = $125
- 10 ISBNs = $295 (best for most authors)
- 100 ISBNs = $575 (best for small publishers)

Recommendation:
Purchase at least 10 ISBNs upfront. You'll need them for:
- Paperback
- eBook
- Hardcover (optional)
- Future editions or books

Step 3: Assign an ISBN to Your Book

Once you've purchased ISBNs, go to "My Account > My ISBNs" and select "Assign Title" for any unused ISBN.

You'll be prompted to enter the following details (based on our real book entry):
- Title: Easy Book Self-Publishing
- Subtitle: A Step-by-Step Guide With AI Assistance
- Author: Nikolay Gul
- Publisher/Imprint: Future-Proof Publishing Press
- Language: English
- Medium: Print
- Format: Paperback
- Format Detail: Trade Paperback (U.S.)
- Publication Date: April 3, 2025 (eBook), April 15, 2025 (Paperback)
- Target Audience: Trade
- Rights Country & Territory: United States, Worldwide
- Title Status: Active
- Is Title Returnable: No
- Price Availability: Available
- Price: $14.99 paperback, $4.99 eBook

- Distributor: Amazon KDP, IngramSpark, Barnes & Noble Press (optional)
- Size: 6 x 9 inches
- Page Count: 130–150 (to be finalized during formatting)
- Illustrations: To be updated (includes images and QR codes)
- **BISAC Codes:**
  - LAN020000 – Publishing
  - COM060000 – Artificial Intelligence
  - BUS043000 – Marketing

Cover Image: Optional but recommended. Use JPG format.

Step 4: Common Pitfalls (and How to Avoid Them)
Bowker's platform is outdated—but functional. Stay patient, double-check fields, and keep your metadata consistent.

Examples of common errors:
- Medium vs. Format Confusion:
  - For eBooks: Medium = E-Book; Format = EPUB or PDF
  - For print: Medium = Print Book; Format = Paperback or Hardcover
- Metadata Mismatch:
  - Inconsistent titles, author names, or formats across Bowker, KDP, IngramSpark, or B&N Press can cause listing or distribution errors.
- **Price Left Blank:**
  - You can skip pricing or write "Write for Info," but it's better to include a real price—even if temporary.

Step 5: ISBN Tips, Tricks & Metadata Best Practices
- Use one ISBN per format (no exceptions)
- Always match title, subtitle, author, and publisher exactly across platforms
- Keep a saved copy of your metadata in a Word doc or spreadsheet
- Assign your ISBN early—even before your book is finished
- Only leave metadata fields blank if they are explicitly marked optional

- Upload your cover later if needed—don't delay the registration just because the design isn't finalized

AI Prompts That Made Our Process Easier
These real prompts helped us complete the process faster and more strategically:
- *Suggest a professional subtitle for my book about self-publishing with AI*
- *Write a keyword-rich book description that sounds confident, not overhyped*
- *Help me create an author biography that includes recent AI publishing achievements*
- *What are the 3 best BISAC codes for a book on publishing, AI, and marketing?*
- *What's the ideal price for my 140-page nonfiction paperback in the business/AI category?*
- *Suggest a clean, punchy front cover slogan for a step-by-step publishing guide*

Final Thoughts: This Is Your Publishing Identity
Your ISBN is more than just a number - it's your book's official fingerprint in the global publishing world. It's how bookstores, libraries, distributors, and readers find and validate your work.

Once you complete ISBN registration the right way, you're no longer "just" a writer. You're a professional, published author with global distribution rights and ownership.

And with AI by your side, the process doesn't have to be slow, confusing, or expensive. You now know exactly what to do—and how to do it right the first time.

Pro Tip: *Even if you choose Amazon's free ISBN, you can't use it outside of Amazon. If you plan to use IngramSpark or B&N later, start with a Bowker ISBN from the beginning—even if it's just for one version.*

Chapter 7: Learn How to Register Your Book with the Library of Congress (LCCN Request Guide for Self-Publishers)

Registering your book with the Library of Congress (LOC) and receiving a Library of Congress Control Number (LCCN) is an essential step in establishing your book's legitimacy, discoverability, and availability for cataloging by national libraries and academic institutions.

This step-by-step chapter is designed to guide first-time self-publishers as well as returning authors (publishing their second, third, or later books).

It also includes real-life insights based on the successful registration process for our first title, "AI-Driven Cybersecurity and High-Tech Marketing" by Nikolay Gul, including tips for sending your deposit copy to LOC the smartest and most professional way.

Step 1: Understand What LCCN Is (and Is Not)
- The Library of Congress Control Number (LCCN) is a unique identifier assigned to your book before it is published.
- It helps libraries catalog your book and makes it easier for them to order and list your title.
- The LCCN is not the same as Copyright registration. These are separate steps in your publishing journey.

Step 2: When Should You Apply?
- Before your book is published.
- Ideally, after you have an ISBN and book title finalized, and before it goes live on Amazon, IngramSpark, or Barnes & Noble.
- Your book must be intended for U.S. publication to qualify.

Quick Clarifications (2025 Updates)

You do not need to upload your manuscript or book file to apply for an LCCN. Only book metadata is required.

LCCN is assigned only to print books (paperback, hardcover). eBooks and electronic-only versions are not eligible. You must already have your Print ISBN before applying and enter it in the form.

Make sure your Publisher Name and City of Publication match exactly with your ISBN registration and copyright page.

Quick Start: Library of Congress (LCCN) Registration as a Self-Publisher:

1. This is the correct starting link (as of 2025):
https://www.loc.gov/programs/prepub-book-link/about-this-program/

2. For First-Time Self-Publisher Registrations:
Select this option:
 "Authors and Self-publishers"
Click on the "Author Portal" (external link).

Direct link to Author Portal:
https://locexternal.servicenowservices.com/pub

3. If you already have an account and want to register a new book (LCCN):

Updated Direct Portal Link (2025):
The current portal is called **PrePub Book Link**:
https://www.loc.gov/publish/pcn/login/
You can log in and manage all your LCCN requests under the **My Requests** tab.
Go to: https://locexternal.servicenowservices.com/

LIBRARY
LIBRARY
OF CONGRESS

🔍 ☰

« PrePub Book Link
🔗 Share

PROGRAM
PrePub Book Link

Menu ⌄

About PrePub Book Link

[Researchers at work in the Main Reading Room of the Library of Congress, close-up view with laptops, paper notes, and books]

PrePub Book Link (PPBL) was modernized to bring together the CIP and PCN Programs into one consolidated system. The entire look and feel of PPBL has been streamlined to be both user-friendly and more attractive. PPBL improves the functionality of the CIP and PCN applications while preserving the mission, which is to provide quality prepublication data to publishers.

Publishers

CIP Publishers

Please use the Publisher Portal ☐ to submit CIP and LCCN requests.

PCN Publishers

Please use the Publisher Portal ☐ to submit LCCN requests.

Please note: Your old PCN accounts (*i.e.* peb47528) were not migrated to PrePub Book Link, so you will need to create an account before you can submit LCCN requests.

Authors and Self-publishers

Please use the Author Portal ☐ to submit LCCN requests.

Note: If you're having trouble signing in or forgot your password, click "Forgot Password" on the login screen.

Pro Tip:

You don't need to register a new account if you have already created one under the **Author Portal.** The "PCN Publisher" and "CIP Publisher" options are for large publishers, not individual authors.

Step 4: Create Your Account (First-Time Authors Only)

- If you're publishing your first book, you will need to create an account in the Author Portal.
- Use a professional email and double-check all your information (name, address, city of publication).

Example:
Publisher Name: Future-Proof Marketing Press
City of Publication: New York, NY
(Make sure this matches your ISBN record and back-of-title-page info)

Step 5: For Returning Authors — Skip Account Creation

If you already registered a previous book with LOC and have an account:
- Log in here: https://prepubbooklink.loc.gov/authorportal/
- Click "Create New Title" or "Submit New Request"

Tip: We encountered confusion during our first registration because the account we thought we had (from older PCN system) did not transfer to the new "PrePub Book Link" system. If this happens to you, create a new account under the Author Portal.

Step 6: Fill Out the LCCN Request Form

Once inside the **Author Portal**, fill out the required fields:

General Information:
- Title: Full title of your book
- Subtitle: Add if applicable
- Language: English (or your book's primary language)
- Projected Publication Date: (e.g., March 11, 2025)
- Number of Pages: Estimate (e.g., 130)

Contributor Information:
- Add your name as Author
- Use full legal name (e.g., Nikolay Gul)

Title Page Info:
- Upload a PDF of your book's title page or fill in manually
- Publisher Name must match exactly (e.g., Future-Proof Marketing Press)
- City of Publication: New York, NY

Print ISBN:
- Add the **ISBN** you've registered (from **Bowker**)
- You do NOT need to register the ebook version unless it's different

IMPORTANT:
Do **not leave the ISBN field empty.**
Enter your full Print ISBN, including hyphens.
Example: **979-8-9927440-1-9**
If you omit this, your application will be delayed or returned.

Book Summary:
Include a compelling but short summary of your book's purpose, audience, and scope.

Additional Info:
Use this section for anything that might help the cataloging librarian (e.g., "Contains cybersecurity, SaaS, and AI marketing strategies").

Step 7: Submit Your Application
- Review all entries
- Submit your LCCN request

You'll receive a confirmation email within minutes and your LCCN will usually arrive via email within 3–7 business days.

Updated Processing Time (2025):

As of April 2025, the Library of Congress states that LCCN requests may take **up to 15 business days** to process.

However, in most cases, authors receive their LCCN within **5–10 business days**.

Step 8: What To Do After Receiving the LCCN

Add your LCCN to your book's copyright page (title verso page)

  - Example: Library of Congress Control Number: **2025902819** - *AI-Driven Cybersecurity and High-Tech Marketing*

- Update your metadata in Bowker, Amazon KDP, IngramSpark, and B&N Press (if they allow it).

Step 9: Mail Your Deposit Copy to LOC (After Publishing)

Once your paperback is published, LOC requires you to send a physical copy of the book. This is called the "deposit copy."

How to Mail It:
- Wait until you receive your author's copy from Amazon or IngramSpark
- Send 1 or more paperback copies (we sent 3 with a cover letter)
- Use USPS Priority Mail to ensure tracking & quick delivery

Address:
Library of Congress, Preassigned Control Number Program
U.S. Library of Congress
101 Independence Ave SE

Washington, DC 20540-4284

[Your Name]
[Your Address]
[Date]
To Whom It May Concern,
Please find enclosed [1/2/3] copies of my newly published book for inclusion in the Library of Congress collection, per the LCCN submission guidelines.

Title: **AI-Driven Cybersecurity and High-Tech Marketing**
Author: [Author's Name]
ISBN: 9798218612481
LCCN: 2025902819
Publication Date: March 11, 2025

Thank you for your work and your service to U.S. authors.

Sincerely,
[Author's Name]

Final Tips
- Always apply for the LCCN BEFORE publishing (pre-publication requirement)
- Keep records of your submissions & email confirmations
- Use AI prompts to help fill forms faster:

AI Prompt (Customize as needed): *"Help me write a compelling 3-sentence book summary for the Library of Congress based on this title and subtitle: AI-Driven Cybersecurity and High-Tech Marketing: Future-Proof Strategies for Sales, Success, and AI-Powered Growth"*

1. Hidden Perk: Free National Validation of Your Book Most new authors don't realize this: once you get your LCCN, your title is officially recorded in the U.S. government's book catalog.

That's instant *credibility*—no expensive press coverage required.

Use this in your book's back matter, press kit, or Amazon description:

"Cataloged by the Library of Congress – LCCN: **[insert number]**"

2. Easy AI Prompt to Write the Perfect Book Summary

Librarians prefer *neutral*, *clear*, and *benefit-focused* summaries—skip the hype.

AI Prompt to use: *"Write a 2-sentence professional summary of my book [Title] that explains its main topic and why libraries or educators should catalog it."*

Tip: Create a mini checklist and cross-verify *before* hitting submit.

Smart Author Framing: A 2-Sentence Book Summary That Sells

Instead of just describing your book, sell it to the cataloging librarian. Try this **AI-powered formula**: *"This book equips [target reader] with [benefit 1] and [benefit 2], using [method or framework]. Perfect for libraries seeking up-to-date resources on [primary topic]."*

AI Prompt: *"Write a 2-sentence Library of Congress summary for a nonfiction book titled Easy Book Self-Publishing: A Step-by-Step Guide With AI Assistance targeting first-time authors."*

Congratulations! You're now professionally registered with the Library of Congress.

This process not only improves your book's credibility but increases its long-term discoverability in schools, universities, national libraries and major search engines.

Chapter 8: Amazon KDP Publishing Guide – Real-Life Solutions & Industry Secrets

Publishing your first eBook and paperback on Amazon Kindle Direct Publishing (KDP) can be both exciting and challenging. Drawing from our first-hand experience, this chapter shares practical solutions to common issues, essential tips, strategic publishing schedules, and insider knowledge that every author needs.

Initial Setup & Account Creation

1. **Amazon KDP Account**

- Visit kdp.amazon.com.
- Create an account using your regular Amazon credentials or set up a dedicated publishing account.
- Complete tax information immediately to avoid royalty delays.

Manuscript & Cover Formatting

eBook Requirements:

- **File Formats:** DOCX, EPUB (recommended)

- **Front Cover:** JPEG, 2560 x 1600 pixels, 300 dpi

- **Formatting Tools:** Kindle Create, Atticus, Vellum (for simplicity)

Paperback Requirements:

- **File Formats:** PDF (fully formatted and fonts embedded)

- **Cover:** PDF full wrap (front, spine, back)

- **Use Amazon's KDP Cover Calculator:** Enter your trim size (e.g., 6" x 9") and page count for precise cover dimensions.

Our Solution: We encountered embedding font issues in PDFs initially. Resolving this required using the "Save As" function in

Microsoft Word, choosing PDF, clicking "Options," and then selecting "ISO 19005-1 compliant (PDF/A)" and "Embed fonts in the file."

Strategic Publishing Schedule

Recommended Schedule:

- Week 1: Finalize manuscript and cover design
- Week 2: Upload to Amazon KDP for initial review
- Week 3: Order proof copies (important to check physical quality)
- Week 4: Soft-launch Kindle edition and start promotional campaigns

Our Tip: Schedule paperback release about 7-10 days after eBook. This allows you to gather reviews and make last-minute corrections if needed.

ISBN & Barcode Clarification

- **Amazon ISBN:** Free, convenient, but restricted to Amazon
- **Your Own ISBN:** Purchase via Bowker for global distribution and professional branding.

Industry Secret: We found using your own ISBN is critical for long-term credibility, global distribution (IngramSpark, Barnes & Noble), and full publishing control.

Pricing Strategies

- Launch Kindle eBook at a promotional price ($0.99 or $1.99) to boost early rankings.
- After 7-14 days, shift to your long-term price ($4.99 - $9.99) for optimal royalty and market position.
- Paperback pricing: Set competitively based on similar books, factoring in print cost and desired royalty.

Upload and Approval Process

- Carefully select relevant categories and up to 7 keyword slots on Amazon KDP.
- Kindle approval typically within 24-72 hours; paperback may take slightly longer.
- Amazon emails you upon approval or flags issues needing correction.

Common Issues & Fixes:

- **Low-Resolution Images:** Ensure all images are at least 300 dpi.
- **Embedded Fonts:** Use PDF/A compliant setting as mentioned earlier.
- **Cover Sizing:** Use the KDP calculator to avoid common cover rejection issues.

Book Launch & Marketing Insights

- **ARC (Advanced Reader Copy):** Distribute to a trusted group for early reviews.
- **Launch Team:** Create a dedicated group to promote, review, and share your launch.
- **Amazon Ads:** Begin with a low daily budget ($5-$10), focusing initially on auto-targeted campaigns for simplicity.

Effective Marketing Actions:

- Create social media posts highlighting book benefits, reader testimonials, and early reviews.
- Leverage author networks (Facebook groups, LinkedIn connections).
- Consider promotional sites like Freebooksy and BargainBooksy for added visibility.

Real-Life Solutions from Our Publishing Experience

- **Issue:** Conflicting ISBN metadata (Bowker vs. Amazon)

Solution: Ensure all platforms (Bowker, KDP, IngramSpark) have identical metadata to avoid rejections or delays.

- **Issue:** Google Books Account Rejection

Solution: Review external link quantity in Kindle eBooks. Google may reject overly promotional external linking. Keep external links minimal or clearly relevant.

- **Issue:** Unclear publication date requirements

Solution: Coordinate publication dates clearly between LOC (Library of Congress), ISBN registration, and Amazon to maintain consistency and professional cataloging.

AI Prompt Examples for KDP:
- "Generate a compelling Amazon description for my nonfiction book titled '[Your Book Title]' focused on [book topic]."
- "Provide a checklist for preparing my manuscript and cover for a flawless Amazon KDP paperback upload."
- "Suggest optimized Amazon keywords for a nonfiction book targeting [audience]."

Sending Deposit Copies to LOC
- After publishing, order author copies from Amazon.
- Use priority mail and include a clear cover letter.
- Sending multiple copies (2-3) is optional but recommended for ease of cataloging.

Publishing your first book on Amazon KDP is a significant accomplishment, made simpler with practical guidance and insights from real-life experience. Remember, every challenge is solvable, every mistake is correctable, and every book is the beginning of your journey. Embrace the learning process, leverage these strategies, and publish with confidence.

Chapter 9: Barnes & Noble Publishing Guide – Why & How to Publish Directly

Publishing directly through **Barnes & Noble Press (B&N)** rather than via third-party distributors like IngramSpark offers significant strategic advantages. In this chapter, we'll clarify why direct registration is beneficial, detail the complete publishing process with practical, actionable insights, and provide AI-powered prompts for seamless decision-making.

Why Publish Directly with Barnes & Noble Press?

When your book is distributed through IngramSpark, it does appear on Barnes & Noble's website—but you lose direct control and the ability to manage your ISBN or metadata independently on B&N's platform.

Key Reasons to Publish Directly:

- **Direct Control Over Your ISBN:** Once distributed by IngramSpark, you cannot independently list the same ISBN directly with Barnes & Noble. Publishing directly allows independent ISBN control and full management flexibility.
- **Better Royalties & Pricing Control:** Direct publishing may offer better royalties and flexibility in pricing strategies.
- **Real-Time Updates & Promotions:** Ability to update metadata instantly and manage promotions without intermediary delays.
- **Enhanced Discoverability:** Better control over categorization and keyword optimization for Barnes & Noble's audience.
- **Exclusive Marketing Tools:** Access to Barnes & Noble-specific marketing and promotional tools directly.

Pro Tip: Always register and publish directly through Barnes & Noble if you plan substantial marketing or targeted sales efforts on their platform.

Step 1: Initial Account Setup

- Visit https://press.barnesandnoble.com

- Create a vendor account; complete bank and tax information.

- Confirm verification promptly to avoid publishing delays.

Step 2: Choosing Your Format & ISBN

- Select your book format (Paperback or eBook).

- Enter your previously purchased ISBN (via Bowker) clearly.

Important: *Avoid conflicts use a unique ISBN not previously assigned via distribution services like IngramSpark.*

AI Prompt: *"Help me verify if my ISBN is already listed through third-party distribution platforms."*

Step 3: Manuscript Preparation & Upload Requirements

- **Paperback Interior:** PDF file, print-ready with embedded fonts.

- **eBook Interior:** EPUB format (recommended), DOCX accepted.

- **Cover File:** Paperback requires a PDF (full wrap including spine and back). eBook requires JPEG (front cover only), 300 dpi.

AI Prompt: *"Convert my paperback manuscript into a fully embedded PDF suitable for Barnes & Noble paperback printing."*

Step 4: Title, Subtitle & Metadata

- Enter your book's exact title and subtitle.

- Select up to five relevant categories to optimize discoverability.

- Include well-researched keywords.

AI Prompt: *"Suggest five highly relevant categories and ten optimized keywords for a nonfiction book about [Your Book Topic]."*

Step 5: Pricing & Royalties

- Set competitive pricing based on comparable books.

- Consider promotional pricing strategies initially (e.g., discounted pre-orders).

- Confirm royalty percentage clearly visible during setup.

AI Prompt: *"Analyze current bestselling books similar to mine and recommend an optimal pricing strategy for initial launch and long-term sales."*

Step 6: Detailed Book Descriptions & Marketing Insights

- Write a compelling book description tailored to Barnes & Noble readers.

- Enhance description with targeted keywords for SEO.

AI Prompt: *"Generate a Barnes & Noble optimized long description and short compelling summary for my nonfiction book [Book Title]."*

Step 7: Author & Contributor Information

- Fill in accurate author biography and contributor details.

- Highlight author credibility, previous publications, and industry recognition clearly and concisely.

AI Prompt: *"Create a concise, professional author biography highlighting relevant credentials, achievements, and previous publications."*

- Add professional editorial reviews if available (up to five).

- Include select testimonials or reviews from credible sources to build trust and encourage purchase.

AI Prompt: *"Suggest a format for showcasing editorial reviews and author endorsements effectively on my Barnes & Noble book page."*

Strategic Publishing Timeline for Barnes & Noble Press

Recommended Timeline:

Week 1-2: Finalize manuscript and cover designs.

Week 3: Upload manuscript and complete all metadata.

Week 4: Set book for pre-order to gather early interest and reviews.

Week 5: Engage marketing efforts (social media, launch teams, promotions).

Week 6: Official book release.

Industry Insight:
Setting your book on pre-order for at least 2-3 weeks maximizes visibility and initial sales impact.

Real-Life Issues & Our Solutions

ISBN Conflict with IngramSpark:
Solution: Request Barnes & Noble customer support remove the ISBN associated with the IngramSpark distribution to allow direct B&N Press listing. Be clear and concise in communication.

Vendor Verification Errors:

Solution: Complete vendor information meticulously, ensuring accurate tax and banking information. If issues arise, contact customer support promptly to resolve errors manually.

Delayed Book Approval:

Solution: Ensure your files strictly adhere to B&N's requirements (embedded fonts, 300 dpi images, accurate cover dimensions). Check the B&N Press guidelines thoroughly before submission.

Practical AI-Powered Decision-Making Prompts

"Is it more advantageous for my situation to publish directly with Barnes & Noble or through IngramSpark?"

"How can I ensure my metadata and ISBN information remains consistent across Amazon, Barnes & Noble, and IngramSpark?"

"Provide a checklist to ensure my Barnes & Noble book submission meets all formatting requirements and will pass immediate review."

Final Thoughts

Direct publishing with Barnes & Noble Press gives you control, strategic flexibility, and tailored promotional opportunities to enhance your book's success. Embrace this streamlined process, take advantage of targeted AI prompts, and launch confidently.

You've got everything you need. Now, publish with Barnes & Noble directly and take charge of your book's future!

Chapter 10: Comprehensive Guide to Publishing with IngramSpark (Paperback & eBook)

Publishing with IngramSpark expands your book's global reach, enabling widespread distribution to bookstores, libraries, and international markets. This guide will clearly walk you through every critical step for successfully publishing both paperback and eBook formats, addressing key questions like whether paying for their eBook conversion service is worth it.

Why Publish with IngramSpark?

Advantages:

- Access to global distribution, including bookstores, libraries, and universities.
- Extensive printing options (trim sizes, paper quality, binding types).
- Professional publishing credibility.
- Control over pricing, returns, and wholesale discounts.

Step-by-Step Guide for Paperback Publishing

Step 1: Account Setup
- Visit IngramSpark.com.

- Create a free publisher account.

- Fill in your banking and tax information accurately.

AI Prompt: *"Provide a clear checklist for setting up my IngramSpark account to avoid verification delays."*

Step 2: ISBN and Title Information

- Assign your previously purchased ISBN (from Bowker).

- Complete detailed metadata (title, subtitle, author, publisher, description).

Important Tip:
Ensure all metadata matches your ISBN data on Bowker for seamless approval.

Step 3: Manuscript Preparation & Requirements

- **Format:** PDF/X-1a:2001 compliant file (embedded fonts required).
- **Trim Size:** Standard sizes like 6"x9" are recommended.
- **Margins:** Minimum 0.5" on all sides, 0.75" gutter recommended.
- **Image Resolution:** 300 dpi minimum for high-quality print.
- **Bleed:** Include if images or graphics extend to page edges.

AI Prompt: *"Create a properly formatted PDF/X-1a:2001 version of my Word document suitable for IngramSpark printing."*

Step 4: Cover Design Requirements

Format: PDF/X-1a:2001 compliant, CMYK colors.

Includes: Front, back, and accurately sized spine (use IngramSpark's Cover Template Generator).

DPI: 300 dpi for clarity and quality.

Barcode Area: Automatically allocated by IngramSpark.

AI Prompt: *"Calculate exact spine width and full-cover dimensions for my 6x9 paperback of 200 pages."*

Step 5: Pricing, Discounts & Returns

- Wholesale discount typically set at 30-55%.

- Decide return policy: No returns, return and destroy, or return and deliver.

Best Practice: Start conservatively (no returns or return and destroy) to minimize financial risk initially.

AI Prompt: *"Suggest an optimal wholesale discount and return policy for my nonfiction paperback on IngramSpark."*

Step 6: Reviewing Proof Copies

- Order physical proof copies to ensure printing quality and formatting.

- Carefully review spine alignment, cover colors, and text readability.

Industry Tip:
Never skip proof reviews—your reputation hinges on quality presentation.

Publishing Your eBook with IngramSpark
Accepted Formats: EPUB only (strict EPUB 3 compliance).

Conversion from DOCX or PDF must result in EPUB 3 validation (use ePubCheck).

Cover: JPEG, minimum 1400 px width, 300 dpi.

Should You Pay for IngramSpark's eBook Conversion ($70)?

Complexity	Recommended Action	Reasoning
Simple (text-focused)	Use free tools (Calibre, Kindle Create, D2D)	Saves money, easy process
Moderate (few graphics)	Try free tools first, pay if validation fails	Balance of cost-efficiency and quality
Complex (graphics, tables)	Pay for IngramSpark conversion	Ensures professional-quality validation passes

AI Prompt Enhancement: *"Assess my manuscript's complexity— considering text density, images, tables, and graphics—and recommend whether to use free eBook conversion tools or pay $70 for IngramSpark's professional service."*

Pros of Paying:

- Professional-quality eBook formatting guaranteed to pass IngramSpark validation.

- Saves considerable time and technical troubleshooting.

Cons of Paying:

- Adds upfront cost ($70 per title).

- You might achieve similar quality for free or less using other tools.

Industry Experience (Our Advice):

- If your manuscript is straightforward (no complex graphics or special formatting), use free tools like Kindle Create, Calibre, or Draft2Digital for EPUB conversion.

- If your book has complex formatting (tables, graphs, heavy images) or you lack technical skills or time, paying

IngramSpark for professional conversion is a worthwhile investment.

AI Prompt: *"Evaluate if paying $70 for IngramSpark eBook conversion service makes sense for my manuscript's complexity."*

Detailed eBook Upload & Approval Process

1. Prepare EPUB file, ensuring compliance with ePubCheck.

2. Upload clearly defined front cover image (JPEG, 1400 px width, 300 dpi).

3. Clearly complete metadata (title, description, ISBN, pricing).

4. Typically approved within 48-72 hours if files meet standards.

Common EPUB Issues & Solutions:

- **Embedded Fonts:** Ensure fonts are licensed and correctly embedded.

- **Formatting Errors:** Validate with ePubCheck tool before uploading.

AI Prompt: *"Help me validate my EPUB file for compatibility with IngramSpark's strict EPUB 3 standards."*

Strategic Publishing Timeline for IngramSpark

Weeks 1-2: Manuscript & cover finalization, ISBN registration.

Week 3: Upload files, review and correct validation errors.

Week 4: Order physical proofs (paperback).

Weeks 5-6: Finalize pricing, launch marketing strategies, set pre-order if desired.

Week 7: Official launch, distribution goes live globally.

Industry Insight:
Start with IngramSpark distribution at least 1-2 weeks after launching on Amazon KDP to ensure metadata consistency, reviews visibility, and stable rankings on Amazon first.

Real-Life Solutions from Our Publishing Experience
- **Issue:** Low-resolution image warnings.

Solution: Check image resolution before upload, using tools like Adobe Photoshop or free online DPI converters.

- **Issue:** Embedded fonts not recognized.

Solution: Export from Word to PDF/A or PDF/X-1a:2001 formats explicitly.

- **Issue:** ISBN conflict across platforms.

Solution: Clearly document each ISBN allocation to avoid duplication conflicts. Assign new ISBNs for different formats (paperback, eBook).

Practical AI-Powered Decision Prompts

"Decide if IngramSpark or direct retailer platforms like Amazon or Barnes & Noble are better for my publishing goals."

"Provide a comprehensive checklist ensuring my files are ready for upload to IngramSpark."

"Recommend marketing and promotional strategies specifically effective for global distribution through IngramSpark."

IngramSpark Tips, Tricks & Time-Savers (Quick Summary)

Use a Professional Email — Avoid generic addresses during account setup to reduce verification delays.

Match Metadata EXACTLY Across Platforms — Your ISBN, title, subtitle, author, and publisher name must be consistent on Bowker, Amazon KDP, and IngramSpark to avoid file rejections.

Free Tools to Create Ingram-Compliant PDFs:
Atticus or Vellum = easiest interior formatting
Canva + Adobe Acrobat = for converting covers to PDF/X-1a:2001
ConvertTown.com = check image DPI instantly

☐ Save the $70 eBook Conversion Fee If You Can: Use free tools like Calibre or Draft2Digital—unless your book has complex formatting (tables, charts, image-heavy).

☐ Wait 7–10 Days After Amazon KDP Launch before going live on IngramSpark to:
Let your Amazon metadata settle
Gather early reviews
Avoid metadata sync issues

Final Thoughts on Publishing with IngramSpark

IngramSpark offers unmatched global distribution opportunities that significantly enhance your book's discoverability and professional reputation. By clearly understanding requirements, strategically timing your launch, and making informed decisions about services (like eBook conversion), you'll ensure a smooth and rewarding publishing experience.
Trust the process, utilize available tools, and let your message reach readers worldwide.

You've got everything you need now let's publish!

Chapter 11: Google Play Books Publishing Guide – Overcoming Upload Quirks & Maximizing Visibility

Publishing on **Google Play Books** places your title in front of a massive global audience. However, the platform has unique upload requirements and common pitfalls. This chapter addresses all upload nuances, requirements, and provides proactive solutions.

Why Publish on Google Play Books?

Massive Audience: Google Books offers global reach through Google Search integration.
SEO Visibility: Enhanced search engine visibility.
Direct Sales Channel: Clear royalties and comprehensive analytics.

Step-by-Step Google Play Books Publishing

Step 1: Account Setup

Visit Google Play Books Partner Center.
Create or use an existing Google Account.
Fill in payment and tax information immediately for smooth verification.

AI Prompt: *"Generate a checklist to complete my Google Play Books Partner account verification efficiently."*

Step 2: ISBN Registration

Use your own ISBN (from Bowker) for eBook version.
Input your ISBN carefully to avoid metadata conflicts.

AI Prompt: *"Verify that my ISBN isn't conflicting across Amazon, IngramSpark, and Google Books."*

Step 3: Manuscript & Cover Requirements

Interior Format: EPUB preferred, fixed-layout PDF accepted (under 2GB).
Cover: JPEG, TIFF, 300 dpi, at least 1000 px wide.

AI Prompt: *"Convert my manuscript to EPUB compliant with Google Play Books' specific formatting requirements."*

Step 4: Metadata Optimization

Clear and optimized title, subtitle, and description.
Select accurate categories and keywords.

AI Prompt: *"Suggest effective metadata keywords and categories tailored for maximum discoverability on Google Play Books."*

Step 5: Pricing and Promotions
Set competitive pricing (align with Amazon Kindle pricing).
Optionally set temporary promotional pricing.

AI Prompt: *"Provide a strategic pricing recommendation for launching my eBook on Google Play Books."*

Step 6: Understanding Google's Unique Quirks & Solutions
Account Rejection without Clear Reasons:

Solution: Avoid excessive external links, ensure professional content quality, and review Google's strict content policy.

"Price Missing" Bug:

Solution: Set publication date to the current date temporarily, then adjust later after approval.

Final Thoughts on Google Play Books

Publishing with Google Play Books enhances your visibility on one of the world's most prominent search engines. Understanding its unique quirks ensures smooth uploading and distribution.

Chapter 12: Apple Books Publishing Guide – Navigating Upload Quirks & Optimizing Visibility

Publishing with **Apple Books** can significantly boost your visibility and revenue potential in Apple's extensive digital ecosystem. However, Apple Books has very specific technical requirements for uploading. This chapter outlines clearly how to meet Apple's precise standards, with step-by-step guidance and actionable AI prompts.

Why Publish on Apple Books?

- **Access to Apple Users:** Directly reach millions of iOS and Mac users.

- **High-Quality Platform:** Premium positioning and high standards for eBooks.

- **Robust Promotional Tools:** Exclusive opportunities for promotions and features in the Apple ecosystem.

Step-by-Step Apple Books Publishing

Step 1: Choose Your Publishing Platform
- Apple Books directly (books.apple.com) or via Draft2Digital for simplicity.

AI Prompt: *"Evaluate if publishing directly with Apple Books or using Draft2Digital is better for my book and technical capabilities."*

Step 2: Manuscript & EPUB Requirements

Format: EPUB only (must pass strict EPUBCheck validation).

Important: No PDF or DOCX uploads accepted.

AI Prompt: *"Validate and format my manuscript into an EPUB compliant with Apple Books' EPUB Check standard."*

Step 3: Cover Requirements

JPEG or PNG, at least 1400 pixels wide, recommended 300 dpi resolution.

AI Prompt: *"Create a high-resolution Apple Books-compliant JPEG cover from my existing design files."*

Step 4: Metadata & Description

Provide a highly optimized description tailored to Apple's audience. Precise metadata and categories enhance discoverability within Apple Books.

AI Prompt*: "Generate compelling metadata optimized specifically for Apple Books' user search behavior."*

Step 5: Pricing, Royalties & Territory Settings

Competitive pricing aligned closely to your other sales channels.

Clearly set sales territories, worldwide or select specific markets.

AI Prompt: *"Recommend pricing tiers and sales territories optimized for my target market on Apple Books."*

Step 6: Apple Books Unique Quirks & Solutions

Stringent EPUB Validation:

Solution: Always pre-validate your EPUB file using free tools like EPUBCheck before uploading.

Delayed Approval:

Solution: Plan for at least one week's lead time in your publishing schedule to allow Apple's manual review process.

Final Thoughts on Apple Books Publishing

Publishing on Apple Books grants access to a premium audience and numerous promotional tools within Apple's ecosystem. Meeting their stringent technical standards ensures smooth publishing and optimal visibility.

AI-Powered Publishing Prompts (Google Play & Apple Books):

"Help me create an EPUB file that will pass EPUBCheck and get approved by Apple Books immediately."

"Clarify how to troubleshoot a Google Play Books account rejection."

"Give me a comparative analysis of my metadata effectiveness on Apple Books versus Google Play Books."

Bonus Tips for Apple Books Success (Not in the Manual)

Test Your EPUB on Real Devices – Don't rely solely on EPUBCheck or Apple's simulator. Load your EPUB file into the free *Books* app on an actual iPhone or iPad to spot formatting issues like image misalignment or strange line breaks that tools might miss.

Use Short, Smart "Subtitles" in Metadata – Apple's search favors clarity. Even if your book doesn't have a traditional subtitle, add one in the metadata (like "For Busy Creators" or "With ChatGPT Prompts") to improve search visibility and set reader expectations fast.

Apple Promo Hack – Authors who publish consistently (or via Draft2Digital with proper metadata) may get flagged for *Apple's Featured Books* or limited-time promotions. You don't apply—*they scout for quality EPUBs and covers.* Clean files, on-point metadata, and accurate categories raise your chances.

Final Words

Clearly understanding and navigating the specific upload quirks of Google Play Books and Apple Books ensures a professional publishing experience and maximizes your book's market potential across two powerful global platforms. Use these guides as your go-to references for hassle-free, high-quality publishing success.

Chapter 13 Publishers Metadata and Keywords for Maximum Visibility

Publishing your book is just the beginning. For your masterpiece to succeed, readers need to find it. That's where metadata becomes your secret weapon.

This chapter teaches you exactly how to optimize every field—from book title to author bio—and gives you ready-to-use templates, AI prompts, and platform-specific tips so you can stand out in the marketplace.

Step 1: Create an Irresistible Title and Subtitle
Strong Example:
Title: The Easy Self-Publishing Blueprint
Subtitle: A Step-by-Step Guide to Publishing and Promoting Your Book Without the Headaches

Checklist:
- [ ] Includes keywords
- [ ] Clear benefit
- [ ] Emotional trigger
- [ ] Genre-friendly

AI Prompt: *"Give me 10 strong title and subtitle combinations for a [genre] book helping [audience] with [goal]."*

Step 2: Write a High-Converting Book Description
HTML Template (Amazon Only):
Ready to Publish Like a Pro?
<ul><li>Step-by-step publishing roadmap</li><li>Free & paid ISBN options</li><li>Formatting tricks</li></ul>

Editable Fields:
[Insert headline], [List 3-5 benefits], [Problem your book solves]

AI Prompt: *"Write an HTML-formatted, persuasive book description for [genre] about [topic]."*

Step 3: Keyword Optimization That Sells
Examples:
- how to self publish a book step by step
- book marketing guide 2025

AI Prompt: *"List 10 long-tail Amazon keywords with buying intent for a [genre/topic] book."*

Your Turn (Fill All 7):
1.
2.
3.

Step 4: Mastering Book Categories
Amazon: 2 default + request 10 total

Example KDP Support Message:
"Hi, please add my book [Title, ASIN] to: Nonfiction > Writing > Authorship, Business > Marketing > Digital Media."

IngramSpark: Use BISAC codes matching Bowker registration

Step 5: Author Bio Optimization
Example:
Nikolay Gul is a bestselling publishing mentor who helps new authors launch books that rank and sell. Visit 911cybersecurity.com.

AI Prompt: *"Write a bio for a [genre] author helping [audience] get [result]. Include CTA."*

Step 6: Platform Metadata Fields to Know
KDP – 7 keywords, HTML description, subtitle, 2–10 categories
IngramSpark – BISAC code, trim size, returnability
Bowker – Format, audience, number of pages, language
Library of Congress (LCCN) – Publisher name, publication year,

author, edition info
Barnes & Noble Press – Genre, audience, print settings

Step 7: Share Manuscript for AI Assistance

Note: GitHub is a free code-sharing site. Gist lets you store text and share a clean link—great for AI or co-authors.

Option 1 – GitHub Gist:
1. Go to https://gist.github.com
2. Paste your manuscript or chapter
3. Click "Create Public Gist" and share link

Option 2 – Google Drive/Dropbox:
- Upload DOCX or TXT
- Set access to "Anyone with link can view"
- Share with ChatGPT or assistant

Example: Final Metadata Strategy
Title: Easy Book Self-Publishing

Subtitle: A Step-by-Step Guide With AI Assistance
Keywords: self publishing for beginners, how to publish book on Amazon, AI book formatting
HTML Description:
Categories: Nonfiction > Publishing, Business > Marketing > Digital
Author Bio: Nikolay Gul, author of AI-Driven Cybersecurity & Publishing Guides

Step 8: Final Checklist
- [ ] Keyword-rich title/subtitle
- [ ] HTML-enhanced description
- [ ] 7 smart keywords
- [ ] Up to 10 book categories
- [ ] Clear author bio
- [ ] Manuscript uploaded for AI help

Chapter 14: Launch Strategy – From Zero to Bestseller (Step-by-Step)

Publishing your book is only half the journey. The launch determines whether your book reaches readers, collects reviews, and earns you visibility or disappears into the digital void.

In this chapter, you'll get a practical, step-by-step launch plan—with real examples and customizable AI prompts to tailor for your topic, industry, or niche.

Step 1: Set Your Launch Date
- Recommended Lead Time: 3-4 weeks from final file upload
- Why: Gives time to gather early readers, reviews, and prepare assets

AI Prompt:
"Create a detailed 4-week book launch timeline for a [topic/genre] nonfiction book targeting [audience type]. Include tasks for reviews, emails, and ads."

Step 2: Build Your Launch Team
- Invite 10–50 beta readers, friends, influencers, or fans
- Give them advance PDF or Kindle copies (marked "review copy")

Checklist:
[ ] Google Form to collect readers
[ ] Private Facebook group or email list
[ ] Early access copy sent
[ ] Review reminder message scheduled

AI Prompt:
"Write an email invitation to beta readers for a [topic] book launch. Make it friendly, include benefits, and a CTA to confirm."

Step 3: Prepare Marketing Materials
These assets should be ready BEFORE your launch:

- Short and long book description
- Author bio
- Review request message
- Social media posts (text and visuals)
- Ads (Amazon/Facebook optional)

AI Prompt:
"Create 5 pre-launch social media post ideas for a [genre/topic] book. Make them engaging, varied, and tailored to [LinkedIn/Instagram/Facebook]."

Step 4: Create A Launch Day Checklist
Day-of Launch To-Dos:
- [ ] Double-check your KDP/IngramSpark settings
- [ ] Post to all social channels
- [ ] Email your launch team
- [ ] Request friends/followers to share
- [ ] Monitor rank & reviews
- [ ] Take screenshots of milestones (e.g., #1 New Release)

Step 5: Follow-Up Plan (Week 2–4 After Launch)
- Ask for honest reviews (again)
- Run discounted promotions
- Submit to book promotion sites (Freebooksy, BookBub, etc.)
- Re-post testimonials
- Pitch guest posts or podcast interviews

AI Prompt: *"Write a friendly follow-up message asking early readers to leave a review on Amazon for a [genre/topic] book."*

Step 6: Optional Advanced Tactics
- Run Amazon Ads (start with $5–10/day)
- Enroll in Kindle Unlimited (if exclusive)
- Apply for Amazon A+ Content (author dashboard)
- Build an email list for your next book

Step 7: Measure Success (and optimize)

Track weekly:
- Reviews gained
- Rank movement
- Conversion rates from ads
- Emails collected

AI Prompt:
"List 10 metrics I should track to measure success after launching a [type of book] in the [topic or niche] space."

Sample Launch Timeline (Nonfiction Example)
Week 1: Final manuscript, start building launch team
Week 2: Send review copies, collect testimonials
Week 3: Schedule social posts, prep email campaign
Week 4: Launch day: post, email, screenshot milestones
Week 5–6: Request reviews again, pitch podcasts, run promotions

Measure Success

Metric	Tool / Source	Goal
Amazon Rank	Amazon KDP Dashboard	Top 10 in niche
Reviews	Amazon + Goodreads	10+ within 2 weeks
Ad Click-Through Rate	Amazon Ads Manager	>0.5% CTR
Email Signups	Landing page / opt-in form	Build list for next launch

Final Tip:
Keep promoting your book long after launch day. A slow burn with consistent effort often outperforms a 1-day spike.

"The best launch teams are built early. Always over-invite. If you want 10 reviews, aim for 30 early readers."

Chapter 15: Book Reviews & Credibility – Secrets to Building Trust Fast

Reviews are the lifeblood of a successful book launch. They're your social proof, your reputation builder, and your credibility magnet. In this chapter, you'll learn how to get more high-quality reviews—fast—using proven strategies, smart tools, and AI-powered messaging.

Step 1: Understand Why Reviews Matter

- Amazon's algorithm favors books with frequent, recent, and verified reviews.
- Readers are 3x more likely to buy books with 10+ reviews.
- Reviews increase discoverability, boost trust, and reduce buyer hesitation.

Step 2: Build Your Review Strategy Before Launch

Checklist:

[ ] Set review goals (10+ within first week, 25+ within first month)
[ ] Identify your beta readers, ARC team, or followers
[ ] Prepare your email/message templates
[ ] Choose where to direct reviewers (Amazon, Goodreads, B&N)

AI Prompt: *"Create a review collection strategy for a [genre/topic] book. Include email drafts, outreach timing, and where to ask for reviews."*

Step 3: Leverage Your Network (Without Begging)

Tips:

- Reach out to friends, colleagues, LinkedIn connections, and readers with a genuine request.
- Don't say "please give me 5 stars." Ask for "an honest review."
- Offer value: "This book is for people like you—your feedback would mean a lot."

Sample Message Template:
Subject: Quick favor?
Hi [Name], I just launched my book on [Topic]! If you get a moment

to leave an honest review on Amazon, it would be a huge help. Here's the link: [Your Review Link]. Thank you so much!

Step 4: Make It Easy To Leave A Review
- Send a direct link to your Amazon review page
  Example: https://www.amazon.com/review/create-review?asin=[Your ASIN]
- Offer a step-by-step image or 3-sentence guide
- Avoid attaching files or asking for logins

AI Prompt: *"Write a short step-by-step guide for how to leave a verified review for a Kindle book."*

Step 5: Keep Following Up – Without Spamming
Smart Follow-Up Sequence:
Day 2: "Hope you enjoyed the book—would love your feedback!"
Day 5: "Reminder: Still time to review and help more readers find this book."
Day 10: "Thanks again! Even 1 sentence makes a difference."

AI Prompt: *"Write a 3-part follow-up email series asking early readers of a [genre] book for a review. Make it friendly and non-pushy."*

Step 6: Use Industry Secrets To Scale
- Include review request at the end of your book
- Ask in your email signature ("Have you reviewed my book?")
- Mention it on social media with a screenshot of an existing review
- Feature reviews in your book description or Amazon A+ Content

Advanced Tip:
Use BookSirens, StoryOrigin, or Booksprout to connect with pre-vetted reviewers.

Step 7: Handle Negative Reviews With Class

- Don't reply emotionally.
- Use feedback to improve your book or future editions.
- Drown negativity in more positive reviews.

AI Prompt: *"Write a professional response to a 3-star review that mentions minor formatting issues but compliments the content."*

Tip: Always place a short, friendly review request at the end of your eBook and paperback. Example: *"If you enjoyed this book, would you take 30 seconds to leave a short Amazon review? It really helps."*

Bonus Tip – In-Person Ask Script:

"If you liked the book, I'd love if you could leave a short Amazon review. It really helps other readers find it. Just search for the title and click 'Write a Review'—super quick!"

Final Thought:

Reviews aren't just vanity metrics. They're the trust bridge between you and your future reader. Start building it early, often, and ethically.

Chapter 16: Beyond Amazon – Multi-Platform Publishing for Maximum Reach

Amazon KDP is a powerful launchpad—but it's not the only destination. In this chapter, you'll learn how to expand your reach by publishing across multiple platforms strategically, without duplication conflicts or legal headaches.

Step 1: Understand Why Going Wide Matters

- Diversifies your income streams
- Reaches non-Amazon audiences (libraries, bookstores, academia)
- Reduces dependency on one ecosystem
- Enhances discoverability through regional and niche platforms

AI Prompt:

"List the top 10 platforms (beyond Amazon) where a nonfiction self-help book can be published or distributed."

Step 2: Know Your Options (Top Non-Amazon Platforms)

1. IngramSpark – Print distribution to bookstores, libraries, academic institutions
2. Barnes & Noble Press – Ideal for U.S. retail visibility
3. Apple Books – Reaches iOS users directly
4. Kobo Writing Life – Dominant in Canada and international markets
5. Google Play Books – Android readers, easy integration with Google accounts
6. Draft2Digital– Aggregates distribution to 10+ platforms in one place
7. Lulu – Custom printing, global distribution, spiral-bound support
8. **Smashwords** – Large eBook catalog, merges with Draft2Digital
9. **BookBaby** – Full-service POD and eBook publishing
10. Scribd – Subscription-based service, ideal for long-term discoverability

Step 3: Format For Multi-Platform Publishing

Print Format Standards:
- Trim size: 6x9 inches (universal)
- PDF file: High-res interior + embedded fonts
- Cover: Use IngramSpark or Draft2Digital's templates (not just Amazon's)

eBook Format Standards:
- EPUB is universal (KDP, Apple, Kobo, etc.)
- Keep layout responsive and simple (avoid fixed formatting)
- Use .docx or upload directly to Draft2Digital for clean conversion

AI Prompt:
"Give me a print and eBook formatting checklist for publishing on both KDP and IngramSpark."

Step 4: Isbn Management

- Use your own ISBN if you plan to publish wide
- Each format (paperback, hardcover, eBook) needs a separate ISBN
- Don't use Amazon's free ISBN if you plan to use IngramSpark (it causes duplication issues)

AI Prompt:
"Explain ISBN strategy for a self-published author using Amazon KDP, IngramSpark, and Draft2Digital."

Step 5: Registration Requirements For Each Platform

KDP – Account: Yes | ISBN: Optional | Format: .docx/.epub | Free: Yes

IngramSpark – Account: Yes | ISBN: Required | Format: PDF | Free: No ($49)

Barnes & Noble – Account: Yes | ISBN: Optional | Format: .docx/.epub | Free: Yes

Apple Books – Account: Yes (Mac only) | ISBN: Required | Format: EPUB | Free: Yes

Google Play Books – Account: Yes | ISBN: Required | Format: EPUB | Free: Yes

Step 6: Strategic Order Of Publication
Recommended Order:
1. Amazon KDP – for fastest launch and early data
2. IngramSpark – for libraries, bookstores (use your own ISBN)
3. Kobo + Apple + Google Play – via Draft2Digital or individually
4. Bonus: CreatePrintBooks or BookBaby for hardcover/spiral-bound options

AI Prompt: *"Create a multi-platform publishing plan and timeline for a self-published nonfiction book using KDP, IngramSpark, and Draft2Digital."*

Step 7: Monitor & Manage Multi-Platform Success
Use a simple spreadsheet to track:
- Upload dates
- Versions
- ISBNs used
- Royalty settings
- Login info

Tips:
- Avoid double-publishing the same ISBN on platforms that cross-distribute (e.g., don't upload to both Smashwords and Draft2Digital)
- Use a consistent book title/subtitle across all platforms
- Monitor royalties and rank individually per dashboard

Final Thought:
Going wide gives you independence and international impact. With the right strategy, your book becomes available anywhere readers are looking—not just on Amazon.

Chapter 17: Copyright, Legal Protections & Your Publishing Rights

Your words are your intellectual property. But without proper registration, protection, and rights management, you may lose control over how your work is used—or worse, watch it get stolen. In this chapter, we walk through the essential legal steps every self-published author must take.

Step 1: Understanding Your Rights

As the author, you automatically own the copyright to your original work the moment you write it. But for legal power and international protection, registration matters.

AI Prompt:
"Explain the difference between owning copyright automatically and registering it formally in the United States for a [genre/topic] book."

Step 2: Registering your Copyright in the U.S.

U.S. Copyright Office Process:
- Go to copyright.gov
- Register as an individual or business
- Upload your manuscript as a digital deposit
- Pay $45–$65 depending on application type
- Receive a certificate of registration in the mail

Timeline: 3–6 months (or pay for expedited service)

AI Prompt:
"Give me a step-by-step breakdown of how to register a copyright for a self-published nonfiction book in the U.S."

Step 3: Using The Right Copyright Page Text

Example Copyright Page:

Optional Additions:
- ISBN
- Edition info
- Cover designer credit
- Editor credit
- Publisher imprint or logo

AI Prompt:
"Generate a complete copyright page template for a nonfiction book about [topic]."

Step 4: Registering With The Library Of Congress
Benefits of Getting an LCCN:
- Your book can be cataloged in libraries
- Increases professional credibility

Free via PrePub Book Link: https://www.loc.gov/publish/pcn/

Requirements:
- Must have your ISBN
- Apply before the book is published

AI Prompt:
"Write step-by-step instructions for how to register a self-published book for an LCCN number in the U.S."

Step 5: Barcode & Isbn Rights Clarity
- Your ISBN identifies you as the publisher
- Never reuse the same ISBN across formats
- You own the ISBN only if you purchase it (e.g., via Bowker in the U.S.)
- Barcodes are generated automatically during publishing or with Bowker tools

Step 6: Contracts, Licenses & Rights Management

If you:

- Hire a cover designer: use a Work-for-Hire agreement
- Use co-authors: agree on profit splits and creative control
- Translate your book: license foreign rights specifically
- Turn your book into a course or audiobook: retain all adaptation rights

AI Prompt:

"List the types of contracts or legal agreements needed when co-authoring, translating, or adapting a nonfiction book for online courses."

Legal Documents Every Author Should Have

A simple reference table can help clarify what's required:

Purpose	Document or Action Needed
Claim ownership	Copyright registration (USCO)
Sell/distribute book	ISBN & barcode (Bowker)
Protect design contributions	Work-for-hire agreement (cover, layout)
Collaborate with others	Co-author agreement or partnership terms
Translate or adapt content	Rights licensing agreement (per region)
Prevent plagiarism	Google Alerts + DMCA takedown readiness

Step 7: Dealing With Piracy & Infringement

What to do:

- Use Google Alerts or Copyscape to monitor for plagiarism
- Issue a DMCA takedown notice if you find a pirated version
- Register your copyright beforehand for legal action and statutory damages

AI Prompt:

"Write a sample DMCA takedown request email for a self-published book found on a pirate website."

Quick Breakdown:

Copyright protects the content (text/design)

ISBN identifies the book's format + publisher

LCCN helps libraries catalog your book

Optional DMCA Email Template (Formatted for Copy-Paste)

Sample DMCA Takedown Request

Subject: DMCA Takedown Request

Dear [Web Host or Admin],
I am the author of the book "[Book Title]," and I have discovered that my work is being distributed illegally on your platform at the following URL: [Insert Link].
This constitutes copyright infringement under the DMCA.
Please remove the infringing content immediately.

Sincerely,
[Your Full Name]
[Your Contact Email]
[Link to your official version, e.g., Amazon]

Final Thoughts:
Legal protection isn't just about preventing theft, it's about retaining the freedom to grow, license, and profit from your intellectual work. Don't skip this step. Claim what's yours.

Chapter 18: Marketing That Works – Proven Strategies to Promote Your Book

Marketing doesn't have to be expensive, complicated, or overwhelming. In this chapter, you'll discover effective, budget-friendly strategies that get your book seen, shared, and sold—plus customizable AI prompts to generate your own campaigns.

Step 1: Define Your Book's Unique Value

Ask:
- What problem does this book solve?
- Who is the ideal reader?
- What emotional benefit does it deliver?

AI Prompt:

"Create a marketing hook and positioning statement for a nonfiction book about [topic] targeting [audience]."

Example:
"This book helps overwhelmed solopreneurs create passive income through self-publishing—without tech skills or a big budget."

Step 2: Build A Strong Author Platform

Assets to Develop:
- Author website with blog + book pages
- Newsletter opt-in with lead magnet
- Social media profiles (LinkedIn, Instagram, or TikTok based on audience)
- Amazon Author Central profile

AI Prompt:

"Create a checklist for building a basic author platform to promote a [genre/topic] book."

Step 3: Leverage Organic & Free Marketing

- Post value-driven content on social media regularly
- Join genre-relevant Facebook/LinkedIn groups

- Pitch guest blog posts or podcast interviews
- Share writing journey and behind-the-scenes updates

AI Prompt:
"List 10 free ways to market a self-published book about [topic] using social media and author communities."

Step 4: Use The "3p" Launch Campaign Formula
P1: Pre-Launch
- Build interest and tease content
- Collect email signups + ARCs
- Ask for shares and pre-orders

P2: Publish
- Launch with coordinated email + post blast
- Submit to BookBub, Freebooksy, Reddit, etc.
- Go live on Amazon Ads or Facebook Ads (low-budget)

P3: Post-Launch
- Share testimonials and milestones
- Offer limited discounts or bundle bonuses
- Invite reader photos and reviews

AI Prompt:
"Design a 3-phase launch campaign for a nonfiction business book including email, ads, and organic posts."

Step 5: Invest In Smart, Small-Budget Ads
Start with:
- Amazon Ads (auto-targeting or keyword-based, $5–10/day)
- Facebook Lead Ads for email collection
- BookBub pay-per-click ads for older books

AI Prompt:
"Write a $10/day Amazon ad strategy for a [genre/topic] nonfiction book optimized for conversion."

Step 6: Use Review & Promo Platforms
Top Options:
- BookSirens (ARC review distribution)
- StoryOrigin (review swaps, email swaps)
- Freebooksy / BargainBooksy (paid promo blasts)
- Goodreads giveaways (builds visibility)

Step 7: Collaborate & Build Authority
- Partner with influencers or thought leaders
- Co-host a webinar or mini event
- Offer bonus content or checklists in exchange for shares
- Create YouTube shorts or TikToks highlighting key ideas

AI Prompt: *"Brainstorm 10 collaborative marketing ideas for a [genre/topic] book to grow visibility and authority."*

Step 8: Track Results & Refine
Metrics to Track:
- Book rank + category position
- Ad spend vs. conversions
- Newsletter subscriber growth
- Social engagement and shares
- Organic mentions or media hits

Tip: Marketing isn't a one-time event it's a rhythm. When you consistently share value, your book becomes part of the conversation.

Ad Platform Comparison

Platform	Budget Range	Best For	Difficulty
Amazon Ads	$5–10/day	Book sales	Easy
Facebook	$5–15/day	Email leads, traffic	Medium
BookBub PPC	$10+/day	Backlist promo, brand	Harder
Goodreads	Giveaway-only	Awareness	Easy

Chapter 19: Turning Your Book into a Brand – Long-Term Growth Strategy

Publishing your book is the beginning, not the end. The smartest authors treat their book like a brand—a foundation for long-term impact, multiple income streams, and deeper influence in their niche. In this chapter, you'll learn how to transform your single title into an author brand, content ecosystem, and revenue-generating machine.

Step 1: Think Like a Publisher, Not Just an Author

Ask yourself:
- What problems does this book solve?
- Who else needs this solution?
- How else can I deliver this content (beyond the book)?

Your book is your intellectual property. It can be adapted into:
- An online course
- A podcast or YouTube channel
- A coaching program or consulting offer
- A series (trilogy, workbook, journal, etc.)
- A keynote speech or TEDx pitch

AI Prompt:

"List 5 ways to repurpose the content of a [genre/topic] nonfiction book into paid or free offers for a [target audience]."

Step 2: Build a Content Ecosystem Around the Book

Create content pillars from your chapters. Examples:
- Each chapter = blog post
- A quote = daily social media post
- A step-by-step from the book = carousel, YouTube short, TikTok, newsletter

Content Ecosystem Funnel Example:
1. Social post (value)
2. Lead magnet (free checklist or chapter)

3. Email list (nurture)
4. Offer (course, consulting, affiliate)

AI Prompt: *"Create a content repurposing funnel for a book about [topic] to attract and convert [ideal audience]."*

Step 3: Develop a Signature Framework or Methodology
People remember frameworks. Turn your content into a branded system.

Examples:
- 5-Step Blueprint
- RISE Method (Reach, Impact, Structure, Execute)
- The B.R.A.N.D. Formula

Name it. Own it. Teach it.

AI Prompt: *"Help me create a branded framework based on the chapters in my [genre] book. Include a 3–5 step model with easy-to-remember terms."*

Step 4: Monetize Beyond Book Sales
Most authors don't get rich from royalties—but they do from what the book leads to.

Monetization ideas:
- Companion workbooks or journals
- Audiobooks
- Paid newsletter
- Affiliate products or sponsor deals
- Paid speaking gigs
- Corporate bulk sales or licensing

AI Prompt: *"Suggest 7 ways to monetize a [genre] book for long-term income outside of Amazon sales."*

Step 5: Scale Into a Book Series or Author Business
A book series = more shelf space + algorithm advantage.

Series ideas:
- Expand on different angles of your niche
- Write deeper guides for specific audiences
- Add case study editions or "Pro" versions

Next-Level Author Business Ideas:
- Start an indie publishing brand
- Offer done-for-you services (ghostwriting, formatting, etc.)
- Run workshops or retreats

Step 6: Protect & Package Your Brand Assets
As your brand grows:
- Trademark your framework or method
- Reserve domain names for your series
- Register social handles before someone else does
- Collect testimonials, stats, and media mentions for credibility

Step 7: Position Yourself as an Expert in the Market
Your book gives you authority. Now it's time to use it:
- Speak on podcasts and webinars
- Publish on Medium, Substack, and LinkedIn
- Apply for awards, book fairs, or media appearances
- Partner with influencers and thought leaders in your space

AI Prompt: *"Write a cold outreach message to invite a podcast host to feature me as a guest based on my book about [topic]."*

Final Thought:
Your book isn't the finish line. It's the foundation. When you think long-term and position your book as the centerpiece of a bigger brand and mission, your readers turn into fans, your content becomes currency, and your publishing success turns into a sustainable business.

Chapter 20: Scaling with AI – Smart Automation for Authors

Imagine having a full-time assistant that never sleeps, helps you write faster, markets smarter, and gives you ideas you never thought of—AI makes that possible. In this chapter, you'll learn how to use AI tools like ChatGPT and others to scale your writing, publishing, and promotion without burnout or hiring a team.

Step 1: Treat AI as a Creative Collaborator, Not a Robot
AI won't replace you—but it can augment you.

Let AI:
- Brainstorm titles, chapter outlines, and book ideas
- Refine marketing copy, blurbs, and ads
- Draft emails, social posts, or course scripts
- Help edit for tone, flow, and clarity
- Generate launch checklists and keyword lists

AI Prompt: *"Suggest 10 engaging chapter titles for a nonfiction book about [topic] targeting [audience]."*

Step 2: Use AI to Accelerate Your Writing Process
Try this fast workflow:
1. Prompt ChatGPT to outline your chapter
2. Flesh it out with your personal stories and voice
3. Use AI to rephrase clunky sentences or summarize long ones
4. Run a grammar/polish pass for tone or flow improvement

AI Prompt: *"Rewrite this paragraph to sound more friendly, persuasive, and authoritative: [paste your paragraph]."*

Step 3: Automate Your Book Marketing with AI
Let AI generate:
- Email sequences (welcome, launch, nurture)
- Social post ideas (hooks, captions, hashtags)
- Landing page headlines and bullet points

- Blurb versions for different platforms
- Reader personas and audience targeting

AI Prompt: "Create 5 different blurbs (100 words each) for my nonfiction book about [topic] to use on Amazon, BookBub, and Facebook Ads."

Step 4: Build a Personalized AI Prompt Library
Keep a Notion, Google Doc, or ClickUp file of your best-performing prompts. Categorize them by:
- Writing & Outlining
- Marketing & Promotion
- Email & Social
- SEO & Keywords
- Metadata & Amazon

Update your prompt vault monthly!

Bonus AI Prompt: *"Give me a full book launch plan with daily actions, suggested content, and hashtags for a [genre/topic] nonfiction title."*

Step 5: AI-Powered Tools Worth Exploring
Besides ChatGPT, explore:
- Jasper.ai – high-quality copywriting
- Sudowrite – fiction-focused brainstorming
- Canva + Magic Write – designs with AI captions
- Descript – podcast/video editing & transcription
- Grammarly + Hemingway – readability, polish
- PromptBase / FlowGPT – prebuilt prompt templates

Step 6: AI for Keyword Research & Metadata
Use AI to:
- Suggest 7 Amazon keywords
- Find profitable book categories

- Optimize subtitles with SEO language
- Compare two versions of a description for readability

AI Prompt: *"Suggest the best Amazon KDP keywords and categories for a nonfiction book about [topic] targeting [audience]."*

Step 7: Stay Ethical & Authentic
AI should support—not replace—your authentic voice.

Guidelines:
- Always inject your personal experience
- Use AI as a collaborator, not ghostwriter
- Credit AI tools where appropriate (especially in acknowledgments or blog content)
- Respect reader trust: they want you, not a robot

Build an AI Assistant That Knows You
Train your favorite AI (e.g., ChatGPT with custom instructions) to sound like *you*, make decisions like *you*, and remember your book's themes. **Feed it:**
Your author bio and brand tone
2–3 past blog posts or chapters
Your book's value proposition and reader pain points

AI Prompt: *"From now on, reply as my author assistant. Learn my tone: **[paste excerpt]**. My audience is **[audience]**. My book helps them **[outcome]**. Confirm you understand before generating."*

Final Thought:
The future of publishing is AI-assisted. Authors who embrace it early will move faster, publish smarter, and build stronger brands without burning out. The tools are here. The intelligence is scalable. What you bring is the vision.

Chapter 21: Book Publishing FAQs – Real Answers for Real Authors

After helping hundreds of aspiring authors navigate the world of self-publishing, we've identified the most frequently asked (and most important) questions authors need answered—clearly and confidently. This chapter delivers fast, actionable insights in a Q&A format.

Q1: Do I really need to buy my own ISBN?

Short answer: If you're planning to publish on multiple platforms like Amazon and IngramSpark, yes.

Why: Free ISBNs from Amazon can only be used on Amazon. Buying your own ISBN from Bowker in the U.S. gives you publishing control, better credibility, and multi-platform compatibility.

Pro Tip: Buy 10 or 100 ISBNs in bulk to save money in the long run.

Q2: Should I use Kindle Unlimited or go "wide"?

Kindle Unlimited (KDP Select):
- Ideal for fiction and fast readers
- Requires 90-day Amazon eBook exclusivity
- You get paid per page read

Wide Publishing (KDP + others):
- Best for nonfiction and global reach
- You control pricing and distribution

Best of Both: Launch in KU for 90 days, then go wide.

Q3: What's the best print-on-demand service for quality and global reach

- Amazon KDP: Fast, free, easy setup. Great for U.S. and eBooks.
- IngramSpark: Better bookstore distribution, hardcover options, and global print quality.
- Use both: Start with KDP, then use IngramSpark with your own ISBN.

Q4: Do I need to copyright my book?

You already own the copyright automatically once your book is written.

But: To legally protect it in court or if it gets plagiarized, register it through the U.S. Copyright Office (copyright.gov). It costs $45–65.

Pro Tip: Register before publishing if possible.

Q5: What file formats do I need for print and eBook?

Platform	eBook Format	Print Format
KDP	.docx or .epub	PDF (with bleed)
IngramSpark	EPUB (Reedsy preferred)	PDF (Ingram specs)

Always preview files in Kindle Previewer and IngramSpark's print simulator.

Q6: How do I get early reviews if I haven't launched yet?
- Build a launch team: Friends, fans, beta readers
- Give them free ARCs (Advanced Reader Copies)
- Ask them to post honest reviews on launch day

Tools to use: BookSirens, StoryOrigin, Prolific, private Facebook groups

Q7: What keywords and categories should I choose?
Use keyword tools like:
- Publisher Rocket
- ChatGPT (with prompts)
- Amazon search bar suggestions

Pro Tip: Pick less competitive categories to rank faster.

AI Prompt: *"Suggest 7 high-traffic, low-competition keywords and 5 KDP categories for a [topic] nonfiction book."*

Q8: How do I price my book for profit and discoverability?
eBooks:
- $2.99–$9.99 = 70% royalty
- Test at $4.99 or $5.99 if nonfiction

Print:
- Consider print cost + royalties
- Price slightly below top competitors

Example Strategy: Launch at $0.99 (promo), raise to $4.99 after 5–7 days.

Q9: What's the most common mistake new authors make?
- Rushing the launch
- Skipping professional editing
- Designing their own cover (without skill)
- Forgetting to collect emails

Remember: Good books fail without good presentation and promotion.

Q10: How can I keep the momentum going after launch?
- Create content around your chapters
- Add your book to your email signature and profiles
- Run ads, do giveaways, update your categories
- Launch a journal, second edition, or mini-course

AI Prompt: *"Suggest a 30-day post-launch marketing calendar for a nonfiction book on [your topic]."*

Publishing your book may start with questions but finishing with clarity is the sign of a confident author. Keep asking. Keep iterating. And use your curiosity as fuel to build a long-lasting career.

Chapter 22: Mastering Book Updates, Relaunches & Second Editions

Publishing your book isn't a one-and-done event. The most successful authors treat their books as evolving assets. Whether it's fixing typos, updating stats, or rebranding for a new audience, updating your book is your secret to long-term relevance, more sales, and SEO dominance.

Step 1: Know When to Update Your Book
Update your book when:
- You've found typos or formatting issues
- You want to add new stats, case studies, tools
- You've updated your branding, niche, or message
- You want to refresh the cover or subtitle for SEO

Pro Tip: Minor changes = update the current version
Major structural changes = release as a Second Edition

Step 2: What You Can Update (Without Republishing)
Interior file (KDP, IngramSpark, Draft2Digital)
Book description and subtitle
Keywords and categories
Author bio
Price

You cannot change:
- Author name (unless you create a new version)
- Book title or ISBN (must publish as a new edition)

Step 3: Turn Updates Into Marketing Opportunities
Don't just quietly update—announce it!

Relaunch Ideas:
- "Now Updated for 2025!" badge on cover
- Run a new promo price or giveaway
- Submit to BookBub/Freebooksy again
- Email your list: "What's New in the Updated Edition?"

AI Prompt:
"Write an email to relaunch the updated edition of my book [title] and highlight what's new and why readers should revisit it."

Step 4: How to Publish a Second Edition (Step-by-Step)

- Create a new manuscript file
- Add "Second Edition" on title page and copyright page
- Purchase a new ISBN
- Upload as a new book on KDP (DO NOT overwrite old listing)
- Optionally unpublish old edition—or keep both live

Bonus Tip: Link your old and new editions using your Amazon Author Central dashboard

AI Prompt: *"Create a publishing checklist for releasing a second edition of my nonfiction book about [topic]."*

Step 5: Use AI to Help Improve the Next Version
Use ChatGPT to:
- Analyze reader reviews and identify improvement areas
- Suggest new chapter ideas based on trends
- Summarize existing content into new formats (courses, audiobooks)
- Rewrite content in a more readable, modern voice

AI Prompt: *"Read these Amazon reviews and suggest 3 improvements for the next edition of my book: [paste reviews]."*

Step 6: Bonus Relaunch Content Toolkit (Use AI for Fast Results)
Social Post Template:
"Big news! The new edition of **[Book Title]** is now live—with updated strategies, fresh content, and bonus resources. Perfect for **[audience]**."

Email CTA Generator **AI Prompt:** *"Write a high-converting email CTA to announce a relaunch for my updated nonfiction book about [topic]."*

Subtitle Refresh Prompt:
"Suggest 5 optimized subtitles for the second edition of my nonfiction book about [problem solved]. Make them catchy and SEO-friendly."

Step 7: Don't Just Relaunch—Repurpose
Use your updated content to:
- Launch a new lead magnet
- Record a YouTube walkthrough or podcast about "what changed"
- Release a bundled "Author's Masterpack" (original + second edition + workbook)

AI Prompt: *"List 7 ways to repurpose the content from a book second edition for more visibility and profit."*

Pro Author Move: Create a "Living Edition" Landing Page
If you update your book frequently—add a private bonus page on your website (e.g., yourbooksite.com/updates). Mention it inside the book ("See updates & bonuses: yourbooksite.com/updates"). Readers will come back, opt in to your list, and even help suggest future improvements. You've just turned one book into an ongoing community.

Smart AI Trick: Use GPT to Find "Update Gaps" You don't need to guess what to update—just ask AI to help. Feed your book's old table of contents or chapter summaries into ChatGPT with this AI prompt: *"Review this list of chapters and suggest any that feel outdated or missing based on 2025 trends in [your book's niche]."*

Final Thought:

Your book isn't static, it's a living asset. Updating it keeps your ideas alive, relevant, and profitable. A smart relaunch can bring in new readers, re-engage your audience, and revitalize your author brand. The authors who grow are the ones who evolve.

Chapter 23: The Ultimate Book Publishing Checklist

There's a reason pilots, surgeons, and NASA engineers use checklists: to avoid fatal mistakes. Publishing your book may not be life-or-death—but it is mission critical. This chapter gives you the Ultimate Book Publishing Checklist, built from years of real-world author experience. Use it to stay calm, confident, and totally in control.

Stage 1: Pre-Writing Setup
- [ ] Book topic and working title defined
- [ ] Ideal reader profile created
- [ ] Competitive analysis done (Amazon, Goodreads, etc.)
- [ ] Outline completed
- [ ] Writing schedule planned

Stage 2: Manuscript Completion
- [ ] Manuscript completed and self-edited
- [ ] Sent to beta readers for feedback
- [ ] Professional editor hired (or trusted AI editor used)
- [ ] Formatting style selected (print/eBook)
- [ ] Final manuscript exported in clean DOCX or PDF format

AI Prompt:
"Create a pre-launch manuscript checklist for a nonfiction book aimed at [audience] about [topic]."

Stage 3: Design & Branding
- [ ] Book cover designed (professional or via Canva/99designs)
- [ ] Title, subtitle, and author name confirmed
- [ ] ISBNs assigned correctly for each format
- [ ] Copyright page finalized
- [ ] Interior formatting tested with previewer tools

Stage 4: Metadata Optimization
- [ ] Book description written (HTML version for Amazon)
- [ ] 7 keyword slots filled (Amazon KDP)
- [ ] 2–10 categories selected and added via KDP support

- [ ] Author bio written and added to all platforms
- [ ] Price strategy determined (promo + full price)

AI Prompt:

"Optimize my book description and subtitle for Amazon SEO for a book on [topic]."

Stage 5: Platform Setup

- [ ] Amazon KDP account created
- [ ] IngramSpark account set up
- [ ] Draft2Digital or Smashwords set up (if needed)
- [ ] Amazon Author Central profile filled out
- [ ] Test uploads completed to all platforms

Stage 6: Launch Marketing Plan

- [ ] Email list built or activated
- [ ] ARC team recruited (10–50 readers)
- [ ] Social media content planned (quotes, reels, carousels)
- [ ] Press release or blog announcement written
- [ ] Book promo sites scheduled (Freebooksy, BookBub, etc.)

Stage 7: Launch Week

- [ ] Book is live on all platforms
- [ ] Launch announcement sent to email list
- [ ] Social media campaign launched
- [ ] Ads (Amazon/Facebook) go live
- [ ] Reviews requested from ARC team

Bonus Tip: Use BookFunnel, StoryOrigin, or BookSirens for ARC delivery and tracking.

Stage 8: Post-Launch Promotion

- [ ] Share Amazon ranking milestones
- [ ] Post 1–2 times per week for next 30 days
- [ ] Pitch 3+ podcasts or guest blogs

- [ ] Submit for awards, giveaways, and contests
- [ ] Prepare bonus lead magnet or workbook

AI Prompt:
"Write a 30-day post-launch marketing calendar for a book on [topic] targeting [audience]."

Stage 9: Long-Term Growth
- [ ] Create companion content (course, newsletter, webinar)
- [ ] Track monthly sales, reviews, ad ROI
- [ ] Repurpose content into reels, carousels, newsletters
- [ ] Plan second edition or follow-up book
- [ ] Update metadata every 3–6 months

Final "Before You Hit Publish" Checklist
- [ ] Everything spell-checked and proofed
- [ ] File tested in Kindle Previewer and print proof ordered
- [ ] Final backup saved to cloud + external drive
- [ ] ISBNs and copyright page double-checked
- [ ] Your author brand assets (website, email, socials) are all aligned and linked

Final Thought:
You don't need to be perfect—you just need to be prepared. Checklists reduce stress, prevent mistakes, and let you focus on what really matters: your message and your readers.
Publishing is no longer overwhelming. It's just one checked box at a time.

Chapter 24: Your Author Brand Toolkit – Stand Out, Sell More, Scale Smarter

A book doesn't sell itself—and neither does an invisible author. If you want your book to have a long shelf life and your message to spread, you need to treat yourself like a brand. In this chapter, you'll build your Author Brand Toolkit—everything you need to stand out and connect with readers, influencers, and opportunities.

Step 1: Define Your Author Identity
Ask yourself:
- What do I want to be known for?
- Who do I serve?
- What's my tone or personality?

Example:
"I'm a practical, no-fluff nonfiction author helping entrepreneurs publish income-generating books without tech overwhelm."

Brand Identity Checklist:
- [ ] Your author name/pen name
- [ ] Bio (short + long)
- [ ] Personal tagline
- [ ] Brand tone (e.g., bold, warm, funny, intellectual)
- [ ] Color palette or visual aesthetic (for website, cover, social)
- [ ] Headshot or avatar image

AI Prompt:
"Write a 3-sentence author bio for a [genre] writer who helps [audience] achieve [result] in a [tone] voice."

Step 2: Create Essential Brand Assets
These are non-negotiables for serious authors:

1. Author Website
   - Book pages
   - Free resource (lead magnet)

- Blog or "What I Write About" section
- Email opt-in

2. Amazon Author Central Profile
   - Photo, links, author story
   - Blog or social feeds

3. Social Media Presence (Pick 1–2 to start)
   - LinkedIn, Instagram, Facebook, TikTok, YouTube Shorts

AI Prompt:
"Create a one-page author website layout for a nonfiction author with one book and a free resource."

Step 3: Positioning Through Content
People don't just buy books—they buy beliefs, energy, and stories.
Share these types of content:
- [ ] Behind-the-scenes (writing, formatting, launch prep)
- [ ] Personal origin story
- [ ] Chapter snippets or quotes
- [ ] Reader reviews and testimonials
- [ ] Lessons learned from publishing

Content Series AI Prompt:
"Generate 10 post ideas for a nonfiction author who wants to grow an audience of [audience] by sharing stories from their book journey."

Step 4: Design a Lead Magnet That Attracts
Offer a freebie that relates to your book and solves a small problem.
Examples:
- Checklist
- Quiz
- Swipe file
- Bonus chapter
- Workbook

AI Prompt: *"Brainstorm 5 lead magnet ideas based on a nonfiction book about [topic] that will attract [ideal audience]."*

Step 5: Email Welcome Sequence Template (3-part)
Email 1: "Welcome + your free gift"
Email 2: "My story and why I wrote this book"
Email 3: "Here's what's inside the book + CTA to buy or review"

AI Prompt Generator: *"Write a 3-part welcome email sequence for an author of a nonfiction book on [topic] who helps [audience]."*

Step 6: Create an Author Media Kit
- 1-line bio + extended bio
- Headshot
- Book cover + back blurb
- Sample interview questions
- Press quotes or testimonials
- Contact + social links

AI Prompt: *"Create a podcast pitch one-sheet for a nonfiction author who helps [audience] using their new book on [topic]."*

Step 7: Maintain a Brand Calendar
Create a simple monthly plan with:
- Weekly blog/newsletter
- 2–3 social posts per week
- Monthly promotion or podcast pitch
- Quarterly relaunch, event, or bonus content

Bonus: Personal Brand Toolkit Prompts
1. Voice Tester:
"Write this blurb in a voice that sounds warm, bold, and helpful: [your paragraph]."

2. Tagline Creator:
"Generate 5 catchy personal taglines for a nonfiction author helping [audience] with [result]."

3. Platform Optimizer:
"List 3 ways to grow visibility for a first-time author with one book and no email list."

Want to look like a bestselling brand even before you are one?

1. Own Your Signature Line or Framework

Create a personal tagline or name your book method (e.g., "The EASY Method" or "Self-Publishing Without Overwhelm"). Memorable language builds trust—and makes your book easier to market, license, and scale.

2. Create an Author Link Hub

Instead of juggling multiple links, build one clean page with your book, lead magnet, and media kit. Use free tools like Canva, Notion, or Linktree. Make it easy for readers, podcasters, and journalists to access everything about you—instantly.

3. Get Found with Google Stacking

Dominate search results by publishing your author story on LinkedIn, Medium, and your own site. Add your book to Goodreads and Google Books. When someone Googles your name or book, you should own the first page.

4. **Consider Trademarking Your Brand**

If your book series, author name, or method has future value, protect it early. Authors who trademark gain long-term credibility—and more leverage in partnerships, media, and licensing deals.

Final Thought:

You're not just an author. You're a brand. A voice. A guide. The more clearly you package that voice, the more people you'll attract—and the more opportunities will come to you.

Your words matter. Now let's make your presence match their power.

Chapter 25: Resources, Tools & Templates for Self-Publishing Success

You're not alone on this journey. Behind every successful author is a toolkit of time-saving, cost-cutting, and creativity-enhancing resources. This chapter shares the best tools, platforms, and templates we personally recommend to help you write, publish, market, and scale smarter—not harder.

Writing & Editing Tools

- Google Docs – Real-time collaboration and easy backups
- Grammarly – AI grammar checker with tone suggestions
- Hemingway App – Simplify and clarify your writing
- ChatGPT / Claude.ai / Gemini – Brainstorming, outlining, rewriting
- Sudowrite – Creative AI for fiction authors

AI Prompt:

"Help me rewrite this paragraph to make it sound friendlier and more confident: [paste your paragraph]."

Formatting & Design Tools

- Atticus – Easiest all-in-one formatting tool for eBook + print
- Vellum (Mac) – Beautiful formatting, drag-and-drop simplicity
- Canva Pro – DIY covers, lead magnets, and social media graphics
- BookBrush – Book mockups, 3D covers, and promotional images
- Adobe Express / VistaCreate – Fast graphic creation alternatives

Template Tip: Use KDP and IngramSpark's free cover + interior size calculators before designing.

ISBN, Copyright & Legal Links

- ISBNs (U.S.): Bowker – myidentifiers.com
- Copyright Registration (U.S.): copyright.gov
- Library of Congress LCCN: loc.gov/publish/pcn/
- DMCA Takedown Template: automattic.com/dmca/

AI Prompt:

"Write a polite but firm DMCA takedown request email for a self-published nonfiction book found on a piracy site."

Publishing Platforms & Distribution
- Amazon KDP: For eBook + paperback
- IngramSpark: For bookstores, libraries, hardcover, global print
- Draft2Digital: Easy wide eBook distribution
- Barnes & Noble Press: Print and eBooks for B&N
- Google Play Books + Apple Books: Direct or via D2D

AI Prompt:

"Create a 5-step plan for uploading my book to Amazon KDP and IngramSpark with matching metadata and ISBNs."

Marketing & Promotion Tools
- BookFunnel: Deliver free or paid eBooks to readers
- StoryOrigin: ARC teams, email swaps, newsletter growth
- BookSirens: Gather early reviews
- Mailerlite / ConvertKit: Email marketing for authors
- Reedsy Discovery: Book reviews and exposure
- Freebooksy / BargainBooksy: Paid promo for discounted books
- Amazon Ads / Facebook Ads Manager

AI Prompt:

"Write a $10/day Amazon Ads campaign plan for a nonfiction book on [topic] targeting [audience]."

Bonus: Templates & Trackers
Create these in Google Sheets, Notion, or ClickUp:
- Launch Checklist Template
- Sales & Ads Tracker
- ARC Team Tracker
- Content Calendar Template
- Metadata Log (Title, Subtitle, ISBN, Categories, Keywords)

AI Prompt: *"Create a Google Sheet layout to track all metadata and version control for multiple publishing platforms."*

Free Tools To Try First

Tool	What It Does	Cost
Reedsy Book Editor	Online book formatting tool	Free
Canva	Graphics and layout design	Free/Paid
Google Forms	ARC and feedback collection	Free
Notion	Dashboard for writing, prompts, content	Free/Paid
ChatGPT (free)	Brainstorming, summaries, rewriting	Free

Pro Author Tip: Build Your Own Toolstack

No two authors use the same tools. Build your stack slowly and prioritize:
- Tools that save you time
- Tools that save you money
- Tools that spark creativity
- Tools that give you visibility

Final Ai Prompt Vault Add-On

Create your own prompt vault by saving your best prompts by:
- Task type (writing, publishing, marketing, etc.)
- Audience/genre
- Goal or intended outcome

AI Prompt: *"Help me organize my best-performing AI prompts into a Notion template or Google Sheet with categories and outcomes."*

Final Thought:

The right tools don't just make publishing easier—they make you faster, sharper, and more empowered. Keep exploring, keep customizing, and keep creating your system. The future author brand is built on both talent and tools tack.

Chapter 26: Final Words & Your Author Journey Forward

You did it. You made it through the entire self-publishing process—from the blank page to a published book, from confusion to clarity, and from idea to income stream.

This isn't the end. This is the beginning of something bigger: your author's journey.

You're Not Just an Author. You're a Creator. A Messenger. A Brand.

Thousands of books are published every day. But yours is different. Why? Because yours has heart. Intention. Purpose. You didn't just publish a book—you built something with meaning. And that has long-term power.

Where Do You Go From Here?

Here's what most successful authors do after launch:
1. **Keep the momentum going**
   - Relaunch with bonuses
   - Run flash sales
   - Add fresh content on social media

2. **Turn your book into other assets**
   - A course
   - A podcast
   - A workshop
   - A paid newsletter

3. **Build a long-term ecosystem**
   - Grow your email list
   - Create companion offers
   - Start writing the next book

AI Prompt: *"Suggest a 90-day content + product strategy for an author of a nonfiction book on [topic] who wants to grow their business and audience."*

Your Book is a Doorway - Not a Destination

The real reward isn't the Amazon rank. It's the email from a stranger who says: "This book helped me. Thank you."
The real success isn't going viral. It's becoming valuable.
And the real fulfillment isn't the sale. It's the impact.

Quick Wins You Can Take Action On Today
- [ ] Schedule a live Q&A about your book
- [ ] Add your book link to every bio and profile
- [ ] Share 1 powerful lesson from your book this week
- [ ] Ask 3 readers to post a photo with your book
- [ ] Submit to 2 podcasts or interviews
- [ ] Create a mini course, bonus PDF, or checklist

AI Prompt: *"Help me repurpose my book into 3 social posts, 1 email, and a video outline for YouTube."*

Author Mindset Reminder

Publishing a book isn't a one-time event. It's a long-term relationship - with your message, your readers, and yourself.
Every chapter you write moves someone else forward.
Every post you share builds trust.
Every prompt you use unlocks more clarity.
Every version you release makes you stronger.

Final Thought

There are two types of people:
- Those who dream of writing a book
- And those who did
You now belong to the second group. That makes you powerful.

Chapter 27: Appendix & Bonus Materials – Everything in One Place

Congratulations you've made it to the final section of the book. This Appendix + Bonus Section is your central command center, giving you direct access to the most essential templates, trackers, prompts, links, and planning tools referenced throughout the book.

Step 1: Universal Book Publishing AI Prompts

Outlining & Writing:
- *"Outline a full nonfiction chapter about [topic] for [audience]."*
- *"Summarize this paragraph and make it more conversational: [paste]."*
- *"Suggest 10 book title + subtitle combos for a book helping [audience] solve [problem]."*

Marketing & Ads:
- *"Write a Facebook ad for a [genre] book targeting [audience] using a bold hook and CTA."*
- *"Generate 5 Amazon keywords for a book about [topic] targeting [audience]."*
- *"Create a 3-part email series for promoting the launch of my book."*

Lead Magnets & Bonuses:
- *"Brainstorm 3 lead magnet ideas for readers of a book about [topic]."*
- *"Write a landing page headline for a free checklist related to [book title]."*

Relaunch & Updates:
- *"Write a launch announcement for the updated edition of my book on [topic] with a CTA and feature list."*

Step 2: Trackers & Templates To Create
Use Google Sheets, Notion, Excel, or ClickUp to build these:

Template Name	What It Tracks
Book Metadata Log	Title, Subtitle, ISBN, Categories, Keywords
Publishing Platform Log	Upload dates, file versions, pricing, ISBNs
Ad Performance Tracker	Campaigns, CTR, CPC, sales, ROI
ARC Review Tracker	Beta readers, contact, feedback, Amazon link
Sales & Royalty Tracker	KDP, IngramSpark, promo spikes, net royalties

AI Prompt:
"Create a Google Sheet layout to track my metadata, ISBNs, and pricing across KDP, IngramSpark, and Draft2Digital."

Step 3: Resource Reference List
WRITING & EDITING:
- Google Docs, Grammarly, Hemingway, Sudowrite, ChatGPT

DESIGN & FORMATTING:
- Canva, Vellum (Mac), Atticus, BookBrush

PUBLISHING PLATFORMS:
- Amazon KDP, IngramSpark, Draft2Digital, B&N Press, Google/Apple Books

LEGAL & ISBN:
- Bowker (myidentifiers.com), copyright.gov, Library of Congress PCN

EMAIL & MARKETING:
- Mailerlite, ConvertKit, StoryOrigin, BookFunnel, BookSirens

PROMO & DISCOVERY:
- Freebooksy, BargainBooksy, Reedsy Discovery, Amazon/Facebook Ads

Step 4: Book Launch Master Checklist (Recap)

1. Final manuscript approved
2. Cover design + spine/back ready
3. ISBNs assigned correctly
4. Book uploaded to KDP / IngramSpark
5. Description, keywords, categories entered
6. ARC readers notified
7. Email list prepped and automated
8. Social media posts scheduled
9. Promo sites confirmed
10. Print proof reviewed
11. Launch day actions executed
12. Reviews collected and tracked
13. Sales, ads, and email stats monitored
14. Bonus content or follow-up plan scheduled

Step 5: Final Reflection Prompt

"Based on everything I've completed so far, what should I prioritize in the next 30, 60, and 90 days to grow as a successful nonfiction author?"

Final Thought:

This appendix isn't the end. It's your dashboard for the next chapter in your career—literally and metaphorically. Come back to it often. Update your templates. Reuse your prompts. Track your growth.

You didn't just publish a book—you created a system for lasting success.

Bonus Chapter 28: Craft Your Unique Writing Style with AI - The Nikolay Gul Method

Why This Chapter Exists

Every bestselling author has one thing in common: a recognizable, authentic writing style. It's what separates bland, forgettable content from writing that feels alive, human, and unforgettable.

But what if you could use AI—not to replace your voice, but to *refine and scale* it?

This chapter shows you how to develop your **own distinct writing voice** using the same approach I use to train AI to write like me: the **Nikolay Gul AI-Style Framework**. It's intuitive, future-proof, psychology-driven, and surprisingly effective.

Whether you're writing nonfiction, fiction, blogs, or marketing copy, this is your blueprint for building a writing style that feels 100% YOU.

Step 1: Understand What Makes Your Style Unique

To personalize your voice, you must first dissect it. Consider these dimensions:

- **Tone** – Casual, formal, witty, bold, compassionate?

- **Rhythm** – Long, poetic flow? Or punchy and snappy?

- **Structure** – Do you use lists, bullet points, short paragraphs?

- **Word Choice** – Do you use slang, alliteration, metaphors, or scientific terms?

- **Emotional Drivers** – Are you humorous, urgent, inspirational, or direct?

- **Industry + Personal Details** – Add niche insights and life experience to stand out.

AI Prompt to Analyze Your Existing Writing Style:

"Analyze the following text and describe its writing style in terms of tone, rhythm, sentence structure, vocabulary, and emotional appeal:

[PASTE 2–3 PARAGRAPHS OF YOUR OWN WRITING HERE]"

Step 2: Create Your Personal Style Blueprint
Now build your own profile for AI to follow and replicate:

Copy + Paste AI Prompt Template:

Going forward, write content in my personalized style with the following rules:

1. Tone: [e.g., practical, optimistic, confident]

2. Rhythm: [e.g., short punchy sentences with frequent line breaks]

3. Structure: [e.g., uses numbered steps, storytelling intros, and strong CTAs]

4. Vocabulary: [e.g., avoids jargon, uses plain language with creative analogies]

5. Emotional Energy: [e.g., inspiring, high-energy, urgency-driven]

6. Personal Details: [Insert 2–3 interesting facts about your background, industry, values, or experience that add credibility and uniqueness.]

Use this style consistently across all future responses.

Pro Tip: Save your blueprint in Notion, Docs, or ChatGPT custom instructions for long-term reuse.

Step 3: Teach AI How to Sound Like You (Without Sounding AI)
AI Prompt for Rewriting in Your Style:

Rewrite the following paragraph in my writing style (defined above) to make it feel more human, persuasive, and emotionally resonant:

[PASTE ANY TEXT HERE]

Use this for:

- Editing blog posts

- Punching up weak copy

- Humanizing AI-generated content

- Drafting emotionally engaging books or ads

Step 4: Use the Psychology of Voice to Deepen Connection
Incorporate these techniques:

- **Open Loops:** Start with a question or tease that compels curiosity.

- **Contrast & Juxtaposition:** "Most books tell you what to do. This one shows you how to do it faster, cheaper, and smarter."

- **Mini-Stories:** Add short, relatable anecdotes (yes, even made-up ones).

- **Emotional Hooks:** Use words that tap into desire, pain, or ambition.

AI Prompt to Create Emotional Hook Intros: *Write 3 emotionally compelling introductions for a chapter about [topic] in my personal writing style that build curiosity, create contrast, and hint at transformation.*

Step 5: Build a Prompt Vault to Reuse Forever

Create categories like:

- **Book Intros**

- **Chapter Hooks**

- **Metadata + Descriptions**

- **Calls to Action**

- **Social Media Posts**

- **Brand Tone Training**

AI Prompt to Generate Your Own Vault Entries:

Using my writing style, create 5 reusable prompts for writing:

1. LinkedIn posts

2. Amazon book descriptions

3. Nonfiction chapter outlines

4. Lead magnets or downloadables

5. Author bios with personality

Save your personal style prompt in ChatGPT's custom instructions or Notion it's your secret weapon for writing that never gets ignored. Don't chase a writing style that sounds like someone else. Create one that sounds like the best version of you—and then use AI to protect, evolve, and amplify it. You already have a voice. This is your toolset to make it unforgettable.

Bonus Chapter 29: The Hidden Power of Human-AI Collaboration

How I Refined My Book's Title, Subtitle & Slogan to Make Readers Say "This is Exactly What I Need"

Why I Questioned My Metadata at the Final Step

As I was preparing to publish **"Easy Book Self-Publishing With AI Assistance Where You Need It,"** I felt confident that everything was ready. The manuscript was complete, the book layout was finalized, the front and back covers were designed, and my ISBN registration was in progress. Yet, right before pressing "Publish," something didn't feel right.

I noticed a repetition in my book's metadata—the title **"Easy Book Self-Publishing"**, subtitle **"With AI Assistance Where You Need It"**, and front cover slogan **"A Step-by-Step Guide to Publishing Success"** all used similar phrases: "Self-Publishing," "Easy," "Step-by-Step." While technically accurate, it felt like overkill. It wasn't persuasive. It wasn't making the reader stop and say: "That's exactly what I need."

That's when I asked myself: Could this metadata work harder for me? Could I make it more emotionally engaging and persuasive without changing the core message?

How AI Helped Validate My Concern

When I shared my concern with ChatGPT, it immediately confirmed what I had sensed intuitively:

Your intuition is 100% correct. The repetition of "self-publishing," "easy," "step-by-step" across Title, Subtitle, and Slogan dilutes the psychological hook.

More importantly, it explained why. The space under the title—the front cover slogan—shouldn't merely echo what's already said. It should deliver an emotional, benefit-driven message that triggers a reader's desire to buy the book, or even gift it to someone else.

The Psychology of a Persuasive Slogan

Together, we outlined what the slogan needed to achieve:

• Be psychologically persuasive
• Be optimistic and helpful
• Focus on outcomes, not process
• Deliver a "That's exactly what I need" moment
• Be gift-worthy
• Highlight practical, ready-to-use benefits

We also leaned on proven persuasion angles:

Empowerment: "You can finally publish your book"
Readiness: "Everything's ready-to-use"
Confidence: "Even if it's your first time"
Urgency: "Don't wait to publish your idea"
Emotion: "Make it real, make it last"

The Final Metadata Refinement

After several rounds of brainstorming, human instinct, and AI-supported refinement, here's what we finalized:

Title:
Easy Book Self-Publishing

Subtitle:
A Step-by-Step Guide With AI Assistance

Front Cover Slogan:
Ready-to-Use Tools, Tips & AI Help to Publish the Book You've Been Thinking About

This slogan is clean, persuasive, and directly taps into the reader's desires:

• They want tools and tips.

• They want AI help without confusion.

• They want to publish the book they've been thinking about (sometimes for years).

This is how AI becomes a co-creator, not a dictator.

Why This Chapter Belongs in This Book

This experience perfectly illustrates the power of human-AI collaboration. It shows that authors should never hesitate to challenge AI, evaluate its suggestions, and refine decisions based on human intuition.

In the end, AI didn't make the decision. I did. But the AI helped me:

• Spot the problem.

• Explain why it mattered.

• Provide improved alternatives.

• Help me clarify my own thinking.

This is how AI can—and should—be used by authors, not only to speed up the publishing process but to make smarter, more confident, and more reader-focused decisions.

Customizable AI Prompts for Authors

You can use these practical prompts to review and refine your own metadata:

1. AI Prompt: Evaluate Metadata Clarity

"Analyze my book title, subtitle, and front cover slogan for repetition, clarity, and emotional impact. Suggest improvements that would attract readers without sounding generic."

2. AI Prompt: Spot Redundancy

"Check my title, subtitle, and slogan together. Highlight if any words or ideas repeat unnecessarily and reduce their impact."

3. AI Prompt: Create Persuasive Slogans

"Generate five short, benefit-driven slogans based on my title and subtitle: [Insert Title & Subtitle]. Make them sound like something readers would want to buy or gift."

4. AI Prompt: Support Human Decision-Making

"Summarize the pros and cons of the title, subtitle, and slogan options we've discussed. Help me decide which combination will make readers think: 'This is exactly what I need.'"

Creativity, Technology & Intuition — At Their Best

This chapter is not just about metadata. It's about what's possible when human creativity, technology, and knowledge work together. The final choice will always be yours. But with AI's help, you can make that decision smarter, faster, and with full confidence.

And now, you've seen it in action - inside the very pages of the book you're holding.

Bonus Chapter 30: Mastering Preorders to Boost Sales and Achieve Bestseller Status

Preordering your book can significantly enhance your publishing success. Top authors regularly use preorders strategically to maximize visibility, drive initial sales, and legally achieve bestseller status.

AI Prompt: *"Analyze my book, '[Your Book Title]', which is a [Genre] novel about [Brief Synopsis]. My primary goals for using preorders are [List 1-3 main goals, e.g., hitting #1 in X category, funding launch ads, building early buzz]. Generate a short (1-2 paragraph) mission statement for my preorder campaign that encapsulates these goals and the book's core appeal for my target audience of [Describe Target Audience]."*

Pros and Cons of Preorders

Pros:

- **Visibility Boost:** Early listings generate excitement, helping your book appear in search results and recommendations well ahead of the release.

- **Bestseller Potential:** Aggregating sales made during the preorder period contributes massively to the first-day sales, pushing your book up the bestseller charts.

- **Marketing Advantage:** Extended lead time allows comprehensive marketing campaigns, advanced reviews, and influencer endorsements.

- **Early Revenue Stream:** Immediate cash flow that can be reinvested into more marketing or publication expenses.

- **Audience Insights:** Analyzing preorder demographics helps fine-tune post-launch promotions.

Cons:

- **Pressure of Deadlines:** Fixed deadlines mean your final manuscript and formatting must be flawless well before the release.

- **Risk of Negative Early Reviews:** Errors or dissatisfaction from early readers can influence overall ratings upon release.

- **Reduced Flexibility:** Changes to book content or metadata post-preorder require additional administrative work and may affect rankings.

AI Prompt: *"Given my situation as a [Debut or Established] author launching '[Your Book Title]' in [Month, Year], with a manuscript readiness level of [e.g., Final Draft, Needs Final Polish, Still Revising] and a marketing budget of [e.g., Small, Medium, Large], analyze the Pros and Cons listed in this chapter. Generate a balanced assessment (bullet points) outlining which factors are most critical for *me* to consider and potential mitigation strategies for the biggest risks (like deadlines or negative reviews)."*

Bestseller Strategies: Proven Techniques for Preorder Success

1. Build Your Preorder Timeline (3-6 months in advance)

Establish your preorder at least 90 days before your release. Announce it alongside a strong marketing push: social media campaigns, email newsletters, and press outreach.

AI Prompt: *"Draft a detailed 90-day preorder timeline template for my book '[Your Book Title]', launching on [Launch Date]. Include key milestones for: final manuscript submission, cover finalization, preorder setup on [Platforms, e.g., KDP, IngramSpark], marketing campaign kickoff (email, social media, ads), launch team mobilization, ARC distribution, reminder pushes, and launch day activities. Structure it week-by-week or by key date ranges."*

2. Leverage Amazon and IngramSpark Algorithms

- Amazon KDP gives extra visibility to books with steady preorder growth.

- IngramSpark allows setting precise preorder periods; leverage discounts strategically to encourage early orders.

AI Prompt: *"My book '[Your Book Title]' is in the [Genre/Category] genre. Brainstorm 5 specific, actionable tactics I can use during the preorder period to potentially improve visibility within Amazon KDP and/or IngramSpark algorithms, focusing on encouraging steady orders and leveraging relevant metadata. Consider strategies related to keywords, categories, pricing adjustments, and early review generation."*

3. Utilize a Launch Team & Early Reviews

- Gather an exclusive launch team to drive reviews immediately on launch day.

- Incentivize early reviewers to preorder by offering bonus content or personal shout-outs.

AI Prompt: *"Generate two customizable message templates:*

1. An enthusiastic invitation email to recruit members for my launch team for '[Your Book Title]', clearly outlining expectations (e.g., read ARC, leave honest review on launch week) and the benefits/incentives offered ([List Your Incentives]).

2. A concise reminder message to send to the launch team a few days before launch, encouraging them to prepare their reviews and thanking them for their support."

4. Clever Marketing & Sales Techniques

- Announce an attractive preorder-only offer (discount, bonus chapter, author Q&A webinar).

- Utilize targeted ads on Facebook, Instagram, LinkedIn, and TikTok to boost preorder visibility.

- Run a limited-time preorder campaign with countdowns and bonuses to create urgency.

AI Prompt: *"Brainstorm 5 unique and compelling preorder-only bonus offers suitable for my book '[Your Book Title]', a [Genre] work targeting [Describe Target Audience]. Go beyond simple discounts and consider exclusive content (deleted scenes, character backstories, resource guides), experiences (Q&A, early access), or limited physical items relevant to the book's theme. For each idea, briefly explain its appeal."*

Publishing Platform Settings for Preorders

Amazon KDP:

- Preorder setup (Kindle and paperback) can be initiated up to one year ahead.

- Ebook preorders are straightforward; paperback preorders recently became more flexible.

- Make use of Amazon's Author Central to enhance author pages during preorder.

IngramSpark:

- Customizable preorder periods.

- Set discounted preorder pricing, and schedule automatic pricing changes post-release.

- Advanced metadata settings allow you to strategically target retailers and libraries.

AI Prompt: *"Generate a detailed checklist of steps required to set up a preorder for an ebook AND a paperback on Amazon KDP for '[Your Book Title]'. Include key decision points like setting the release date, uploading manuscript/cover files **(draft vs final),** pricing, territory rights, category/keyword selection, and enabling preorder* *functionality. Add a similar checklist for IngramSpark, highlighting differences like wholesale discount settings. "*

Use these AI-generated prompt templates for creating effective preorder campaigns:

AI Prompt 1 – Bestseller Status Campaign: "Write a persuasive, compelling social media announcement encouraging immediate book preorders for '**[Your Book Title]**'. Include urgency, exclusive preorder benefits, and clearly communicate the value readers gain from ordering early. Tone should be enthusiastic, friendly, and confident."

AI Prompt 2 – Email Marketing Blast: "Generate a professional, highly engaging email aimed at existing subscribers to secure preorders of '**[Your Book Title]**'. Highlight exclusive preorder content, limited-time offers, and succinctly explain why early ordering will greatly benefit readers."

AI Prompt – Facebook Ads: "Write highly converting Facebook ad copy optimized for preorder sales of the book '**[Your Book Title]**', including a short headline, persuasive description, and call-to-action encouraging immediate action."

Practical Industry Secrets & Tips

- Set up keyword-optimized preorder landing pages to capture early sales.

- Continuously tease preorder content snippets across your social media to maintain engagement.

- Approach influential bloggers and podcasters offering advance copies in exchange for preorder promotions.

AI Prompt: *"Generate creative ideas for the following, tailored to my book '[Your Book Title]' ([Genre/Brief Synopsis]):*

1. Three content snippets I could tease on social media (e.g., a compelling quote, a short atmospheric description, a character introduction).

*2. Two potential angles or hooks to pitch influential bloggers/podcasters in the **[Your Genre/Niche]** space for preorder promotion or ARC reviews."*

Future-Proofing Your Strategy

- Keep track of algorithm updates on Amazon and other platforms to adjust preorder promotions effectively.

- Integrate AI insights regularly to maintain competitive edge— use AI prompt optimization techniques to refine marketing messaging.

Advanced Preorder Strategies & Industry Insights

Here are some additional tactics and deeper insights to maximize your preorder campaign's effectiveness:

A. Platform Nuances & Cross-Platform Strategy:

1. **Leverage All Major Platforms:** Don't just focus on Amazon. Set up preorders directly via Kobo Writing Life, Apple Books, Google Play Books, etc. Each platform has its own charts and visibility algorithms; direct preorders count more heavily there. Use a distributor like Draft2Digital or PublishDrive to easily manage wide preorders if needed, but direct is often best for key platforms.

   AI Prompt: *"My primary sales platform goal for '**[Your Book Title]**' is **[e.g., Maximize Amazon rank, Reach wide audience including libraries, Target Apple Books users]**. Develop a strategic recommendation on which platforms **(KDP, IngramSpark, Kobo, Apple, etc.)** I should prioritize for setting up direct preorders vs. using an aggregator like Draft2Digital, explaining the rationale based on my goal and the advice in* *this chapter."*

2. **IngramSpark for Hardcovers & Library/Bookstore Reach:**
   Use IngramSpark specifically for hardcover preorders **(often unavailable or limited via KDP initially)** and to target libraries and bookstores. Set wholesale discounts strategically (e.g., 55% returnable) during the preorder phase to encourage bulk orders from retailers. Be aware there can be reporting delays from IngramSpark sales compared to KDP.

3. **Amazon's "Also Boughts" Seeding:** Preorders help train Amazon's algorithms. Early orders start associating your book with others bought by the same readers, populating the crucial "Customers who bought this item also bought" section with relevant titles right from launch day, boosting organic discovery.

4. **Category Strategy *Before* Preorder:** Research and select your Amazon KDP categories carefully *before* setting up the preorder. Choose less competitive categories where your preorder numbers have a higher chance of hitting the #1 Bestseller rank within that specific category on launch day. You can change categories later, but initial placement matters for early visibility.

5. **Audiobook Preorders:** If releasing an audiobook, set up preorders through your distributor (e.g., ACX, Findaway Voices/Spotify, Authors Republic). Audiobook preorders contribute to overall launch momentum and have their own charts (e.g., on Audible).

Next-Level Marketing & Engagement Tactics:

6. **Tiered Preorder Bonuses:** Instead of one bonus for everyone, create tiers. Example: "First 50 preorders get a signed bookplate + bonus chapter," "Next 100 get the bonus chapter,"

"All preorders get a digital resource guide." This incentivizes *early* action.

7. **Preorder Swag & Proof of Purchase:** Offer exclusive physical items (bookmarks, stickers, art prints, signed bookplates) for those who submit proof of preorder (a screenshot of their receipt). This builds deeper fan connection but requires managing fulfillment. Use a Google Form or dedicated service for submission.

8. **Cross-Promotions with Author Peers:** Partner with other authors launching books around the same time. Promote each other's preorders to your respective email lists or social media audiences. This expands reach significantly.

9. **Universal Book Links:** Use services like BookLinker, Books2Read, or Geniuslink to create a single link that directs readers to the preorder page on their preferred retailer and in their country. This simplifies promotion and improves conversion rates globally.

10. **Video Trailer Power:** Create a compelling book trailer and feature it prominently on your preorder landing page, in ads, and social media posts. Video can significantly increase engagement and convey the book's tone/promise quickly.

11. **The "Soft Launch" Preorder:** Before the main public announcement, quietly make the preorder available to your core audience (e.g., email list, Patreon supporters). This helps catch any technical glitches, gathers initial orders to build momentum, and makes your superfans feel valued.

AI Prompt: *"Design a compelling tiered preorder bonus structure for '[Your Book Title]'. Define 3 distinct tiers (e.g., Early Bird - first 50, Preorder Crew - next 150, All Preorders) with increasingly valuable, thematic rewards relevant to a [Genre] book. Also, outline the logistical steps needed to manage proof-of-purchase submission (e.g., using Google Forms) and fulfillment for potential physical swag like [Your Swag Idea, e.g., signed bookplates]."*

12. **Clarify "Bestseller" Goals:** Be specific. Preorders are highly effective for achieving Amazon *category* bestseller status **(e.g., #1 in "Technothrillers")**. Hitting major lists like the *New York Times* or *USA Today* typically requires thousands of sales within the first week *including* preorders, often heavily reliant on bookstore sales tracked via BookScan **(requiring strong IngramSpark distribution and retailer relationships).** Manage your own and your audience's expectations.

13. **Dynamic Campaign Adjustments:** Don't "set and forget" your preorder campaign. Monitor which marketing channels (email, specific ad platform, influencer post) are driving the most preorders *during* the campaign. Reallocate your budget and efforts towards what's working best in real-time.

14. **Post-Launch Momentum is Crucial:** Preorders get you a strong start, but sustained sales *after* launch day are vital for maintaining rank and visibility. Plan marketing efforts for launch week *and beyond*. The preorder spike is just the beginning; avoid the "launch day cliff."

AI Prompt: *"Define 3 specific, measurable, achievable, relevant, and time-bound* **(SMART)** *goals for the preorder campaign of '[Your Book Title]'. Include at least one goal related to a specific Amazon category bestseller rank, one related to total preorder numbers, and one related to tracking marketing channel effectiveness (e.g., 'Achieve X* **preorders via email marketing').** *Suggest key metrics to monitor daily/weekly during the campaign."*

Cutting-Edge Preorder Tactics & Refinements

Building on the advanced strategies, here are further refinement and innovative approaches:

1. **Preorder Price Pulsing:** Beyond a single preorder discount, create short-term "flash sale" deeper discounts during the preorder window. Announce these exclusively to your email

list or social media followers for 24-48 hours to create multiple spikes of urgency and reward engaged followers.

2. **Gamify Your Preorder Goals:** Turn hitting preorder milestones into a community event. Create a visual tracker (e.g., a thermometer graphic) on your website or social media. Announce specific rewards unlocked at different preorder levels (e.g., "At 100 preorders, I'll release a deleted scene," "At 250, we unlock character art," "At 500, I'll do a live Q&A"). This encourages collective effort.

3. **Strategic Partnership Diversification:** Look beyond just author cross-promotions. Partner with bloggers, podcasters, influencers, or even small businesses whose audience overlaps with your book's target readers, but who aren't authors. Offer joint giveaways or exclusive content related to preorders.

   AI Prompt: *"Develop a concept for either a 'Preorder Price Pulse' campaign OR a 'Gamified Preorder Goal' campaign for '[Your Book Title]'. Outline the core mechanic, the timeline within the preorder period, the specific offers/rewards, and the communication plan (how/where it will be announced and tracked)."*

4. **Leverage Preorder Period for Ad Testing:** Use the longer lead time to actively A/B test different ad creatives, target audiences, and copy on platforms like Amazon Ads or Facebook Ads. Analyze click-through rates (CTR) and conversion rates *for preorders* to identify the winning combinations *before* your main launch week ad spend.

5. **Maximize KDP Metadata Attributes:** When setting up your book on KDP, delve deep into all available metadata fields, especially the often-overlooked "Attributes" (available for certain categories/genres). Ticking relevant attributes (e.g., specific tropes, character types, settings, themes) can give

Amazon's algorithms more data points to match your book with the right readers.

6. **ISBN Strategy - Own Your Number:** While KDP offers a free ISBN, purchasing your own ISBNs (e.g., via Bowker in the US) provides greater control and flexibility. It designates *you* as the publisher of records across all platforms and formats using that ISBN. This is particularly important for coordinating efforts with IngramSpark and ensuring consistent branding and metadata control, which can subtly impact how sales data aggregates.

7. **Pre-Preorder Hype Campaign:** Don't just announce the preorder when it goes live. Build anticipation *beforehand*. Tease the cover reveal, share intriguing snippets or character introductions, run polls related to the book's themes, and hint at the upcoming preorder announcement and exclusive bonuses. Get people excited *before* they can click "buy."

8. **Explicit ARC Review Disclosure Guidance:** When managing your launch team, explicitly instruct ARC (Advance Reader Copy) recipients that if they choose to leave a review, they *must* include a disclosure like "I received a free copy of this book and am leaving this review voluntarily" to comply with platform policies (like Amazon's) and FTC guidelines. This protects both you and the reviewer.

AI Prompt: *"Generate 3 distinct social media post ideas for a 'Pre-Preorder Hype Campaign' for '[Your Book Title]', designed to build anticipation *before* the preorder link is live **(e.g., cover snippet reveals, character poll, 'ask me anything' about the book)**. Also, provide a concise, compliant draft sentence for ARC reviewers to include in their reviews regarding receiving a free copy."*

A Brief Note on Bestseller Expectations

While preorders can substantially boost your book's visibility and ranking, it's important to set realistic expectations. Becoming a category bestseller on platforms like Amazon is achievable for most dedicated authors through targeted strategies like those described here. However, making it onto major bestseller lists such as *The New York Times* or *USA Today* typically requires thousands of sales within a short window and often involves coordinated retail distribution and substantial marketing investments.

Focus your efforts on clearly defined, achievable goals first such as hitting bestseller status in a specific Amazon category, then build on that success as your readership grows. This measured approach helps manage expectations while still celebrating meaningful, achievable milestones in your publishing journey.

Acknowledgments

To my wife –

Thank you for being the calm in my chaos, the quiet strength behind every crazy idea, and the steady voice of reason when I need it most.

You've supported every project, caught the things I missed, and somehow managed to keep life running while I chased deadlines, brainstorms, and the occasional "just one more tweak"—all while gently reminding me that sleep and meals are not optional.

Your patience, humor, and faith in me—especially on the days I didn't deserve it—mean more than I'll ever be able to explain.

I may be the one making the plans, but you're the reason they keep moving forward—with love, balance, and just enough sarcasm to keep us both sane.

Recommended Resources & Discover More

ISBN & Publishing Information: Official ISBN provider in the U.S. Bowker (MyIdentifiers): myidentifiers.com

Library of Congress Control Number (LCCN): https://loc.gov/publish/pcn

U.S. Copyright Office. Protect your intellectual property officially: https://copyright.gov

Amazon Kindle Direct Publishing (KDP): https://kdp.amazon.com/en_US/help/topic/G201499380

IngramSpark,- Professional distribution for bookstores and libraries globally: https://www.ingramspark.com/lp/author-resources

Draft2Digital Aggregator for multiple eBook platforms (Apple Books, Kobo, etc.): https://draft2digital.com

Apple Books (for Authors). Ideal for Mac/iOS ecosystem publishing: https://authors.apple.com

Google Play Books - Direct integration with Google services: https://play.google.com/books/publish

Marketing & Promotion Resources:

Nikolay Gul book Author: https://www.linkedin.com/in/webdesignerny/

Goodreads Author Program. Connect directly with readers and manage author profile: https://goodreads.com/author/program

BookBub - Powerful promotion platform for discounted and free books: https://bookbub.com/partners

LinkedIn Marketing Solutions. Ideal for professional, business, or nonfiction authors: https://business.linkedin.com/marketing-solutions

Future-Proof Your Publishing Knowledge (AI Prompt): *"Provide me with the latest trends, tools, and resources in self-publishing, marketing, and AI-driven content creation, specifically tailored for nonfiction/business authors."*

If this book helped you even a little, please consider leaving a quick Amazon review. It means the world to indie authors like me—and helps others discover this book too.

You don't need to be perfectly prepared.

You've done something most people only dream of: you turned an idea into a book. That's not just a milestone it's a statement. It means you're ready to share, grow, and lead. This book was designed to make publishing easier, faster, and smarter. But what comes next is even more powerful: expanding your message, growing your audience, and turning your book into long-term impact. So keep going. Keep learning. And remember: every author starts with one book—but the ones who keep building become a brand.

About the Author: Nikolay Gul

Nikolay Gul is a practical self-publishing author dedicated to helping authors launch smarter, faster, and without overwhelm. He developed the unique 'Human Writing Style Framework,' enabling authors and businesses to personalize AI-driven content, ensuring it authentically reflects their own voice and brand for maximum impact. Leveraging extensive experience in AI prompt engineering, Nikolay crafts advanced strategies to optimize AI performance, enhance content quality, and streamline the creative process, as demonstrated throughout this guide. In addition to empowering authors, Nikolay is a marketing strategist known for helping MSPs, high-tech brands, and cybersecurity firms maximize their marketing and sales results. His expertise spans AI applications in branding, sales automation, and content strategy.

Nikolay is also the author of "**AI-Driven Cybersecurity & High-Tech Marketing**" (**ISBN: 9798218612481 | LCCN: 2025902819**), a hands-on guide featuring 100+ AI-powered marketing prompts designed to accelerate business growth through real-world AI applications an approach mirrored in the creation of this book.

Connect with Nikolay:
https://www.linkedin.com/in/webdesignerny/

Easy Book Self-Publishing – A Step-by-Step Guide to Format, Publish, Promote & Sell Your Book Disclaimer:

This book, provides information for educational purposes only. While the author has strived for accuracy, the self-publishing landscape and AI technology change rapidly; therefore, the information herein may not always be current or suitable for your specific situation. Use this information at your own risk.

This book does not constitute professional, financial, or legal advice. Consult qualified professionals for specific guidance. The author and publisher make no guarantees regarding results, income, or success from applying these strategies. AI-generated content examples require careful review and adaptation. The author and publisher disclaim any liability for loss or damage arising from the use of this book or its content.

www.ingramcontent.com/pod-product-compliance
Lightning Source LLC
Chambersburg PA
CBHW071653210326
41597CB00017B/2196